www.brookscole.com

www.brookscole.com is the World Wide Web site for Wadsworth Publishing Company and is your direct source to dozens of online resources.

At www.brookscole.com you can find out about supplements, demonstration software, and student resources. You can also send e-mail to many of our authors and preview new publications and exciting new technologies.

www.brookscole.com
Changing the way the world learns®

Essential Interviewing

A Programmed Approach to Effective Communication

Sixth Edition

David R. Evans
University of Western Ontario

Margaret T. Hearn
University of Western Ontario

Max R. Uhlemann
University of Victoria

Allen E. Ivey
University of Massachusetts

THOMSON

™

BROOKS/COLE

Australia • Canada • Mexico • Singapore • Spain
United Kingdom • United States

THOMSON
™
BROOKS/COLE

Exective Editor: Lisa Gebo
Assistant Editor: Alma Dea Michelena
Editorial Assistant: Shelia Walsh
Technology Project Manager: Barry Connolly
Marketing Manager: Caroline Concilla
Marketing Assistant: Mary Ho
Advertising Project Manager: Tami Strang
Project Manager, Editorial Production: Tom Novack
Print/Media Buyer: Kristine Waller
Permissions Editor: Sarah Harkrader

Production Service: Carlisle Communications, Inc.
Text Designer: Carolyn Deacy
Photo Researcher: Pat Quest
Copy Editor: Sharon O'Donnell
Cover Designer: Roger Knox
Cover Image: Daryl Benson/Masterfile
Cover Printer: Webcom
Compositor: Carlisle Communications, Inc.
Printer: Webcom

For more information about our products, contact us at:
**Thomson Learning Academic Resource Center
1-800-423-0563**

For permission to use material from this text,
contact us by: **Phone: 1-800-730-2214**
Fax: 1-800-730-2215
Web: http://www.thomsonrights.com

Library of Congress Control Number: 2003100354

ISBN 0-534-55848-8

Brooks/Cole—Thomson Learning
10 Davis Drive
Belmont, CA 94002
USA

Asia
Thomson Learning
5 Shenton Way #01-01
UIC Building
Singapore 068808

Australia/New Zealand
Thomson Learning
102 Dodds Street
Southbank, Victoria 3006
Australia

Canada
Nelson
1120 Birchmount Road
Toronto, Ontario M1K 5G4
Canada

Europe/Middle East/Africa
Thomson Learning
High Holborn House
50/51 Bedford Row
London WC1R 4LR
United Kingdom

Latin America
Thomson Learning
Senecca, 53
Colonia Polanco
11560 Mexico D.F.
Mexico

Spain/Portugal
Paraninfo
Calle/Magallanes, 25
28015 Madrid, Spain

Contents

Chapter Three

Effective Questioning   *43*

Chapter Four

Reflecting Content   *61*

Chapter Five

Reflecting Feeling   *79*

Chapter Ten

Information Giving   173

Chapter Eleven

Interpreting   195

Chapter Twelve

Structuring for Information and Action   211

Chapter Thirteen

Enlisting Cooperation 233

Chapter Fourteen

Putting It All Together 261

Additional Resources 289

Index 291

Preface

*T*he publication of the sixth edition of *Essential Interviewing* marks the 25th anniversary of the first publication of this book. We are amazed and humbled that this book has stayed the course over the years. We are also aware of and pleased that many students of all ages have profited from its existence. As in past editions, our primary goal in this edition was to update the content and remain current.

At the heart of this and past editions lies the importance of individual communication skills. A group of core communication skills is essential to any interview, whether it takes place in counseling, nursing, social work, personnel work, or information gathering. *Essential Interviewing* defines these communication skills and demonstrates how to use them effectively in many kinds of interviews.

The single-skills focus and branching programmed structure used in *Essential Interviewing* have proven to be effective in developing communication skills in lay and professional workers. Even those who initially were skeptical about the utility of the programmed-learning model have come to value its efficiency in preparing individuals to communicate effectively in a variety of face-to-face situations.

Competent interviewers, regardless of orientation and degree of training, find that communication is basic to their relationships with clients. Physicians, nurses, ministers, teachers, psychologists, counselors, and other professionals who are constantly involved with other individuals, and paraprofessionals, from crisis workers to community volunteers, must also know how to communicate in order to effect beneficial change. Parents note improvements in their relationships with their children when they exercise the skills discussed in this book. Couples can use the material presented here to improve their relationships with each other. *Essential Interviewing* offers techniques that can be useful in conducting personnel interviews and business conferences and in developing harmonious work relationships.

The sixth edition of *Essential Interviewing* reflects continued feedback from our colleagues who have used previous editions. The first chapter has been revised to reflect current thinking. Chapter 13 titled "Enlisting Cooperation" has been revised to include material on how to manage the impact of stages of change on client cooperation. As necessary, a section outlining the definition and purpose of the skills taught has been added to the beginning of the chapters. As with past editions, teaching frames are clearly identified. At the end of each skill chapter, the section titled "Cultural Considerations" has been revised.

We gave extensive consideration to incorporating material on culture and gender into each chapter. Given the introductory nature of this book, however, we decided that the first priority of this book is mastery of the single skills in their generic form, with supplementary material on their cultural applicability provided at the end of each chapter in the "Cultural Considerations" section. Although the literature provides some information on the use of these skills within some cultures, this is primarily descriptive; there is little empirical evidence of their use within myriad cultures that interviewers may encounter.

Because of its emphasis on developing particular skills, *Essential Interviewing* can be used in self-training or as a basic text for courses and workshops. The mastery of any skill depends on practice. Activity Units, including practice interview checklists, are included at the end of each chapter to facilitate the development and integration of communication skills. Chapter 1 outlines the historical principles of and approach to the acquisition of skills presented in the book. Chapters 2 through 6 focus on building rapport and seeking information—skills that are basic to all effective communication. Chapters 7 through 14 examine specific skills that can help bring about change. Other works that can be used to supplement this text are listed in the "Additional Resources" section at the end of the text.

We would like to express our appreciation to the following reviewers, who shared their students' reactions and suggested improvements to the text: Martin Ritchie, The University of Toledo; Nancy Janus, Eckerd College; James Motiff, Hope College; Virginia David, Nazareth College of Rochester; Barbara McLean, Onondaga Community College.

We also wish to acknowledge the cooperative support of our colleagues and students in our respective communities.

David R. Evans
Margaret T. Hearn
Max R. Uhlemann
Allen E. Ivey

Foreword

*E*ssential Interviewing was originally published 25 years ago. What explains the book's popularity and longevity? Is it the book's content? Methodology? Updated editions? Authors? All of the above? Let's start with the content.

A fundamental skill in humanizing relationships is good communication. A book devoted to improving communication skills will be highly valued by professionals who rely on effective interviewing to facilitate rapport and information gathering. These professionals work in a variety of settings including medical centers and clinics, psychological consulting rooms, classrooms, community agencies, and business corporations. Effective communication is not only a professional concern but also a concern of our communities, families, and religious and social organizations. Professionals, paraprofessionals, and the lay public will find these communication skills important in their work and personal relationships. Thus, the book's content addresses an important professional and social need—the need for effective communication.

Although the content is highly valued, the way content is presented is key. In response to the ambiguity and complexity of effective communication, the authors have adopted a behavioral skills curriculum. Several components contribute to the skills curriculum including skills analysis and a social learning methodology. The skills analysis is an analytical task of decomposing complex phenomena such as communication into the constituent parts of single skills. The microskills strategy was used in education at Stanford University in the 1960s and later applied by Allen Ivey, one of the authors, to counseling. Using Ivey's microcounseling approach, a group of core communication skills were identified based on the pioneering work of Carl Rogers.

The single-skills format defines and structures the heart of the book. Within each skill-focused chapter, a social learning methodology is followed:

skill definition (instructions), examples (models), opportunities for practice (rehearsal), and feedback. Margaret Hearn, another of the book's authors, combined microcounseling and the social learning approach with programmed learning to produce a more efficient way to teach communication skills. Hearn's branching program format, the format of the book, breaks down the microcounseling skills into smaller teaching units that progressively build toward skill acquisition. This microskills-social learning-branching program strategy has been empirically supported and successful through six book editions.

The book has not remained static, but has incorporated current conceptual thinking and empirical findings. For example, the single-skills focus tends to convey a generic application without due regard to differing communication contexts. Although culture-specific communication lacks an established research foundation, the authors have provided a section on cultural considerations within each chapter. The integration of culture and communication is welcomed into our multicultural world that needs culturally competent and aware professionals and communities.

Six editions of any book required knowledgeable and dedicated authors. I have had the privilege of working with each of the authors, including being a faculty colleague in the Psychology Department at the University of Western Ontario (Evans, Hearn, and Uhlemann), professional colleague for many years in the Division of Counseling Psychology of the American Psychological Association (Ivey), and a collaborator on microskills research (Evans, Hearn, and Uhlemann). From such professional and personal perspectives, these authors are uniquely qualified to keep on providing effective and efficient communication training. Mention has already been made about the original contributions of Allen Ivey to microcounseling and Margaret Hearn's branching program tool. Max Uhlemann was an early microskills researcher and teacher and continues his scholarly and practice pursuits. In more recent publications, David Evans and Allen Ivey have demonstrated a commitment to culturally relevant practice in clinical and health psychology and counseling and psychotherapy, respectively. These authors have been given the rare opportunity of crafting a gift over 25 years. They have delivered a significant contribution to the practical art of good communication. Congratulations!

Gerald L. Stone
Professor Emeritus
Counseling Psychology
University of Iowa

Chapter One

Programming a Foundation for Learning

*C*lients come to us with many stories. They may share narratives of their concerns or problems—financial issues, interpersonal conflict, accomplishments, or deep internal stress. This book is about drawing out people's experiences and listening to them. There is great comfort in being heard.

Listening and helping clients express their thoughts and feelings about life experiences serve as a foundation. Then we are prepared to help clients find new meanings in their experiences and, perhaps, even develop new interpretations of what happened.

Finally, clients need to act. Understanding one's experiences may provide a foundation, but action is also required. As an interviewer, counselor, or therapist, often your task is to help clients find new ways of thinking and feeling as well as new behaviors and coping skills.

Essential Interviewing will help you to develop the skills required to:

- Listen and help clients describe their experiences.
- Help clients think about their lives and the world in new ways.
- Assist clients in what needs to be done.

You will find that the listening skills and strategies of *Essential Interviewing* are central to learning how the client's story or problems developed. Once you understand the client's way of thinking and feeling, you are better prepared to use skills such as confronting and interpreting to help them find new meaning in their experiences—enabling them to discover new ways of thinking, feeling, and acting. Three systems provide background for this text.

Narrative theory and constructivism provide theoretical background for this approach (Ivey, D'Andrea, Ivey, & Simek-Morgan, 2002; Kelly, 1955; Monk, Winslade, Crocket, & Epston, 1997; White & Epston, 1992). The systemic program and thoroughly researched framework of the microskills approach (Ivey, 1971; Ivey & Ivey, 2003) provides the technology for taking theory directly into practice. Hearn (1976), one of our authors, combined these ideas into a new and effective approach for counseling and training using programmed learning. Programmed learning takes single microskills of the interview and further breaks them down into manageable teaching units or frames. In programmed learning each teaching frame builds on the preceding one, providing both knowledge and practical experience in the use of each microskill. You will find an example of the programmed-learning approach below as we start to examine how to help clients describe their experiences and learn new ways to cope with their issues and concerns.

Making Decisions during an Interview

Imagine that you are interviewing the woman in the photo for the first time. After some preliminaries and rapport building, she says to you:

1.1

CLIENT:		I'm seeing you because I can't talk to anyone. My lover seems so strange and cold lately. Chris seems so depressed. What should I do?

Choose the statement that best reflects what you might say in response.

INTERVIEWER: Things have been bad for you lately. You're wondering what to do. (Go to 1.2)

INTERVIEWER: What specifically has Chris been doing that gives you that impression? (Go to 1.3)

INTERVIEWER: I think we ought to bring in both you and Chris so we can talk about the relationship in more depth. (Go to 1.4)

| **1.2** | **Your answer:** | *Things have been bad for you lately. You're wondering what to do.* |

Correct. This reflective response suggests to the client that you want to hear more before you provide an answer. It is indicative of a listening style that attempts to find out what is going on with the client before taking action. The reflective response also focuses on the client in front of you. Most often, interviewing is at maximum effectiveness if we give prime attention to the person in front of us.

| **1.3** | **Your answer:** | *What specifically has Chris been doing that gives you that impression?* |

This open question provides you with more data and facts about the situation but is less effective. The focus is on Chris rather than the client in front of you. This response might be useful later and could be helpful in a problem-solving approach to helping in which the goal is to find out as much as possible before making a decision.

| **1.4** | **Your answer:** | *I think we ought to bring in both you and Chris so we can talk about the relationship in more depth.* |

This response places responsibility on the interviewer immediately. Although the idea of couples counseling may be very important, the interviewer provides structure at too early a point. Generally, it is considered best to "listen before you leap."

As you can see in the above example, the programmed approach helps to examine very specifically what is occurring between the interviewer and the client throughout the session. As you may have noted, each of the three responses has some merit and the programmed approach makes it possible to examine the pluses and minuses of each. Throughout this book, we will be emphasizing listening to the client's experiences first—then we can work on client understanding and action. This text cannot supply you with the answers to every interviewing situation, but it does provide you with some basic and important guidelines that ultimately you will integrate into your own style.

An effective interviewer can make a tremendous difference in the life of another human being. Rather than telling you what to do, we want you to participate actively by dealing with the client's story, choosing the appropriate response, and practicing the essential dimensions of interviewing. You may expect to acquire a basic knowledge of the interviewing process, equip yourself with useful skills important in interviewing and in day-to-day interactions, and establish a foundation for further practice as you develop your own unique style of working with others.

Although this text focuses on the client, both interviewer and client grow in the process of effective interviewing. As you read and participate in the step-by-step process of this programmed text, you may encounter ideas and skills you can experiment with in your daily life—skills that can help you understand others and yourself more completely. Moreover, as you develop competence in relating to others, the material presented here will help you understand what skilled interviewers are doing and will enable you to adopt the elements of their styles that seem valuable to you.

This book provides you with specific formulas and methods that have proven useful in a variety of interviewing situations. However, no one method or interviewing skill is appropriate in every situation or with every individual. Therefore, we return to *you*. You, as an interviewer, are the individual who can make a difference if you combine the ideas and skills presented here with your own knowledge and experience. Use this text as a tool: challenge, evaluate, and shape the material, maintaining your individuality and personal genuineness in your relationships with others. Effective use of the tools of interviewing can enhance the development of another human being; poor interviewing can be destructive. If the tools provided in this text become an end in themselves and you act as a slave to them, you will not be an effective interviewer.

Steps toward Clarifying the Interviewing Process

Counseling, psychotherapy, and interviewing have traditionally been seen as complex, almost mystical activities immune to systematic study and definition. However, over the years, several trends have emerged that have made the art of interviewing more definable and specific. The following paragraphs provide a brief historical review, which may be helpful in understanding the background of the skills and concepts presented in *Essential Interviewing*.

The Freudian legacy   Sigmund Freud is the founder of psychotherapy. Before the time of his brilliant theorizing, a general conception of mental functioning did not exist. Freud brought order out of chaos. Although his ideas remain controversial, his influence cannot be denied. Freudian thinking also gradually introduced the idea that it was possible to educate psychotherapists. Psychoanalytic education involved supervision and abstract discussions of the interview and did not consider specifics of what went on between therapist and client in the session—training was based on what the therapist *remembered* from the interview.

The Rogerian revolution   The most influential psychotherapist of modern times is Carl Rogers, who was the first to record his interviews and make them widely available to the profession. The psychoanalytic approach to helper training was unchallenged until Blocksma and Porter in 1947, using the then-new medium of audio recording, discovered that what people *say they do* did not necessarily correspond to *what they actually do* in the interview. The influence of Rogers and this groundbreaking research has forever changed the field.

Rogers's writings are a balance of theory with actual interview transcripts (Rogers, 1942, 1957). Not only did Rogers say what he did but he also provided explanations; as he provided his ideas, he listened to others. His theories constantly changed and expanded during his lifetime, and at the end of his career, he was actively involved in group work, action for peace, and personal power (Rogers, 1977).

Rogers's legacy is a *listening legacy*. More than any other person, he stressed the importance of listening and *hearing* the client's story. A brief look at one aspect of his theory highlights the importance of listening to the client. Figure 1.1 underlines the fact that although the majority of experiences of the interviewer and client are unique, they do share some experiences. For example, they may live in the same city, travel on the same transport system, and have had similar difficulties growing up. However, they may come from different cultures, different age groups, and whereas one may have moved around the country, the other may have remained in his or her city of origin. The listening skills, attending behaviors, open questions, and reflections of content and feeling are important tools to help the interviewer understand the unique experiences of the client.

Carl Rogers

Figure 1.1
The unique and shared experiences of an interviewer and a client.

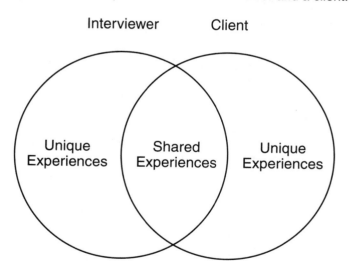

If the interviewer has understood the client's unique experience, then the client will confirm this to the interviewer. Here is an example:

CLIENT: I came home from work last night after a 12-hour shift and my roommate had gone out and left the apartment in such a mess. What made me even more angry was the fact that there were dirty dishes everywhere.

| INTERVIEWER: | It must have been distressing for you to come home last night after your long shift and have to face a messy apartment. |
| CLIENT: | Yes, you're right. I was so exhausted, I thought of leaving and going to a friend's place. |

If the interviewer has failed to understand the client's experience accurately, then the client will often correct, rephrase, or reiterate what was said. Here is an example:

CLIENT:	I came home from work last night after a 12-hour shift and my roommate had gone out and left the apartment in such a mess. What made me even more angry was the fact that there were dirty dishes everywhere.
INTERVIEWER:	It must have been distressing for you to come home last night after your long shift and have to face a messy apartment.
CLIENT:	Not really; my roommate does that all the time. My anger settles down and I just clean up the place.

These examples show how the listening skills and the subsequent responses help the interviewer to come to understand the client's unique experiences. It also underlines the importance of not moving to more action-oriented interviewing phases until the interviewer fully understands the unique experiences of the client.

Microskills and extending the listening legacy One problem in interviewing analysis is that the beginner is often overwhelmed by information because of the complexity of the process. As Rogers (1957) suggested, people seem to learn and remember best those things that are developed step-by-step. The question is how to break down the complex interviewing process and determine which skills are most useful under varying circumstances.

Two research teams in the late 1960s attacked this complex question. One team used videotape to help evaluate counselor–client interaction and identify the specific, observable behaviors that constitute effective interviewing (Ivey, Normington, Miller, Morrill, & Haase, 1968). A second team used a complex statistical approach to interview transcripts (Zimmer & Park, 1967). The findings of the two studies were virtually identical, supporting the idea that it is possible to delineate specific skills of helping. Parallel work by Carkhuff (2000), Egan (2002), Hill and O'Brien (1999), and others also focus on specific skills. Recent writing has extended the skills of this book to group work (Ivey, Pedersen, & Ivey, 2001). Such programs, now widely available, have become standard in the training of interviewers, counselors, and psychotherapists.

| **1.5** | *You have listened to Susan for a few leads, and it may be time to establish a structure for your session. Which of the following responses might you make to Susan? Each of them helps provide a framework for later decision making.* |

SUSAN: Well, I guess I'll admit it. Chris did throw something at me this morning. (tearfully) It was the bread knife. He felt terrible, and so did I. He's gotten terribly angry with me so often. It's not the first time.

INTERVIEWER: Susan, there's a lot going on right now. Let's continue now to explore. But first I'd like to know about Chris and more details about what's happening. Then I'd like to know how all this affects you. Then, perhaps together, we can move to some decision making. (Go to 1.6)

INTERVIEWER: It is important that we decide what we need to work on here. First, we need to keep you safe. Then we need to look at your plans for the future. Later we can work out some of the things you may be feeling. (Go to 1.7)

INTERVIEWER: I hear you loud and clear. But before we take any action, I need to know more about your thoughts and feelings. Tell me more about how all this affects you and how you are feeling right now. (Go to 1.8)

1.6 Your answer: *Susan, there's a lot going on right now. Let's continue now to explore. But first I'd like to know about Chris and more details about what's happening. Then I'd like to know how all this affects you. Then, perhaps together, we can move to some decision making.*

Correct. This structuring response acknowledges the complexity of the session and the need for more data. Although the client needs to explore her feelings, this situation is possibly an emergency. Thus, it is critical to bring out key facts about Chris and the relationship. This can serve as a foundation for learning about Susan's feelings. All these together can serve as a frame for needed decision making. As a first step in decision making, we need to organize all the key data and draw out the client's story.

1.7 Your answer: *It is important that we decide what we need to work on here. First, we need to keep you safe. Then we need to look at your plans for the future. Later we can work out some of the things you may be feeling.*

There are some very good aspects to this response, particularly the egalitarian "we" and the emphasis on safety. At the same time, the helper is taking almost all control and direction, leaving little decisional power in the client's hands. The decisional plan is organized almost solely by the interviewer. We have not heard enough of the narrative to know whether this is a correct decision.

1.8 Your answer: *I hear you loud and clear. But before we take any action, I need to know more about your thoughts and feelings. Tell me more about how all this affects you and how you are feeling right now.*

This is a good listening response and will reinforce more client talk about herself, but it misses important aspects of a potentially critical situation. The reinforcement structure is perhaps too limited and fails to deal adequately with potential violence.

In the above example the programmed approach enables us to examine what is occurring between the interviewer and the client. Each of the three responses takes the interview in a different direction. The programmed approach makes it possible to examine the result of each of these different interviewer responses.

As you can see, many things are happening at once in an interview. It is important that you listen and try to enter the client's frame of reference. Simultaneously, it is vital that you be aware of the client's situation and context and be ready to take action when it is appropriate. Action taken too forcefully and too early may drive a needy client away. Action taken too late may leave the client in a damaging situation.

This all is made more complex by the dynamics of the unique person before you and his or her relationship with others. Despite abuse and violence, it is difficult for women to leave relationships. When they do leave, the average woman tends to return to an abusive relationship for several more tries before she realizes that the situation is hopeless. In situations of potential crisis such as this, your ability to balance listening and action will be challenged constantly.

Implications of the example case and *Essential Interviewing*'s programmed approach   The example interview with our client, Susan, is designed to introduce you to how this book is organized. You will work through representative case material in a step-by-step fashion. Specific skills of interviewing will be stressed as you work through cases similar to those you may encounter.

The intent of this brief section is to provide some examples of how later study in theory on your part will help guide your decisions. This book, however, focuses on commonalities and qualities required for an effective interview. Building on the skills provided in this text, you will be prepared to encounter theories and decide how you personally want to help others grow and develop in a social context.

To meet the different treatment needs of a widely varying clientele, you must develop a wide-ranging response capability. In a single day, you might work with a 72-year-old depressed retired person, a job-hunting teenager, and a Hispanic individual victimized by discrimination, each with different needs.

Out of this confusing array of conflicts, theories of helping, and demands from clients, you must choose the most appropriate intervention. This text develops some beginning ground rules in surviving the interview and provides a foundation for your future growth as you develop increasing interviewing competence and confidence.

Let us now examine how single helping skills can be useful in developing some of the behaviors of the effective interviewer.

Microcounseling and the Single-Skills Approach

This book is based on a systematic method of interviewer, counselor, and therapist training termed *microcounseling* or *microtraining* (Ivey, 1971; Ivey, 2003).

These terms are used because microportions of interviews are identified and classified. Making the interview clear and specific helps beginners (and many advanced professionals) master the complexities of the interview and the counseling and therapy process more quickly and fully.

The microskills approach is the most thoroughly researched interviewing training program with over 450 data-based studies attesting to its effectiveness (see the Research Box on page 14). It was the first interviewing skills training program to emphasize cultural and gender differences. Translated into at least 15 languages, the program has won wide cultural acceptance. You'll find microtraining programs used with social workers in the Canadian Arctic and in Aboriginal Australia. United Nations Educational, Scientific and Cultural Organization (UNESCO) has trained AIDS volunteers in Africa, whereas refugees in Sri Lanka and Uganda have experienced the microskills training systems. The Japanese Microcounseling Association is a center of research and practice and the Netherlands has been the site of extensive research on medical applications (patients do better if physicians listen to them). Eriksson Telecom in Sweden and Standard Electric Lorenz in Germany, as well as business people throughout the world, have profited from training in these microskills. Of course, the center of all this work is the counseling and interviewing field.

The single-skills approach Throughout this book, you will encounter chapters on specific skills of the helping process. Starting with the most basic skill of attending behavior, you will gradually build expertise in foundational inquiry and listening skills to help clients share their experiences. This will be followed by action or influencing skills, which provide you with additional alternatives to help clients act.

The following exercise will help acquaint you with the single-skills approach at a more experiential level. Microcounseling maintains that moving abstract concepts of interviewing to concrete action facilitates the learning process.

Experiencing the single-skills approach Let us imagine that you are to be interviewed by one of the two interviewers in the photos. Which one would you prefer? If you were our client Susan, would you open up as readily to interviewer 2 as would to interviewer 1? A warm, empathic attitude is expressed as much or more by nonverbal behavior as it is by words and your responses as an interviewer.

What specific behaviors do you see in the pictured interviewer? Behaviors are things that can be seen, observed, touched, and counted. "Poor listening" is not a behavior. When searching for behaviors of interviewing, try to make them specific. Lack of eye contact, crossed arms and legs, and leaning back on the couch are specific, observable behaviors that you can observe in interviewer 2. What are the specific, positive behaviors of interviewer 1?

Interviewer 1

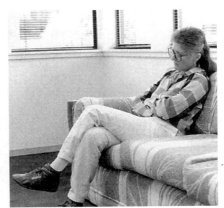

Interviewer 2

Most people in Western society prefer the first interviewer. The active, involved posture is considered a sign of interest and caring. However, some people prefer the second interviewer. Imagine that you are sharing something very difficult with an interviewer. For some, the intensity of interviewer 1 will be threatening and intrusive. Moreover, some cultures (e.g., Native American, Dene, and traditional Hispanic) might consider interviewer 2 more thoughtful and respectful.

Although general principles and skills can be defined, you must always modify them to meet the needs of the person before you. At times, you will need to interpret and reframe the narrative and apply different theories and combinations of skills.

Try This Exercise

As you start this book, it will be helpful if you try the following exercise. It will help make the ideas expressed in this book much clearer and potentially more useful to you over the long run. Find a partner who is willing to be interviewed by you. If at all possible, audiotape the session or have a third person watch the interview and give you feedback. Here are your instructions for the interview:

Imagine you are to conduct an interview; however, you are to do the *worst* job possible; do as many things wrong as you can, and do them deliberately. Be creative and, above all, ineffective in the session. As a result of being ineffective, you can later define some positive aspects of interviewing. Spend about 3 minutes in this interview. Remember, do the worst job possible and exaggerate to make clear what you are doing. After you have completed the exercise, go back and list *specific* things that indicate a poor interviewing technique.

This procedure is similar to the first step of Rogers's innovation in training, in which interviewers listen to high- and low-quality sessions. However, the exercise presented here involves you and asks you to identify in precise terms the *specific* things that are done incorrectly.

In one classroom exercise, the following list was drawn up by the students:

interrupted	maintained poor eye contact
interviewers talked all the time	paid no attention to emotions
listened poorly	contradicted the client
gave extensive advice	appeared bored at times
maintained sloppy posture	expressed no empathy
played with lighter	seemed phony
didn't seem interested	looked at floor frequently

This list describes many inappropriate interviewing behaviors. Obviously, to be effective, you should try to develop skills that counter these behaviors. However, correcting these problems all at once may be too difficult for many beginning interviewers. A more feasible approach is to select just a few behaviors, perhaps only one, master it, and then move on to others. While doing this, it is important to remember that there are many possible narratives and many cultural styles of listening.

Where Are You Heading?

This book is a structured-learning experience in 12 basic-skill areas crucial to all types of interviewing. An overview of these areas, with a brief summary of the key dimensions of each, is provided below. Note that the first four skill areas—attending behavior, effective questioning, reflecting content, and reflecting feeling—emphasize *listening* to the client. Your first task is to draw out the client's experiences.

You may find that, in early practice interviews, you ask one or two questions, obtain a sense of the client's problem, and suggest solutions; for example, "Have you tried talking to your parents?" "Why not try . . . ?" Too many questions followed by quick answers is not effective interviewing. We urge patient and careful attention to these four listening skill areas because, unless you can truly hear the client's concern, there is little chance of effective change. Patient and careful clarification of the client's narrative in detail separates interviewing from the routine superficial advice that friends often give one another.

We strongly suggest that you do not skip ahead to the later chapters of this book until you have mastered these basic introductory skills at a high level, as assessed in Chapter 6, and can effectively conduct an interview with a real or role-played client for at least 5 minutes without using skills that could result in direct behavior change. Once you have demonstrated that you can listen, move on to the skills presented in later chapters. You are most helpful when you give clients time and room to describe their experiences fully.

Theories of Counseling and the Microskills

A frequent question asked of those who use the microskills approach is how the skills relate to varying theories of counseling and therapy. For example, what type of skills does a Rogerian person-centered counselor use? A cognitive–behavioral helper? A psychodynamically oriented professional? An interviewer focused on multicultural issues?

The general answer is that all of the theoretical orientations use microskills, but they tend to use them in different patterns. The key points follow:

All theoretical orientations stress the importance of listening skills Regardless of how they think about counseling and therapy theory, almost all effective interviewers are good listeners. They attend; they tune into and reflect content and feelings; and when appropriate, they summarize what the client has been saying.

Different theories emphasize listening to different things You will find that a Rogerian person-centered helper will pay considerable attention to thoughts and feelings about the self, particularly as these manifest themselves in the interview. The cognitive–behavioral counselor will be interested in listening to specific behaviors the client engages in and how the client thinks about those behaviors. In turn, the psychodynamic therapist listens for past experience and will correspondingly focus on and pay less attention to other issues. The multiculturally aware interviewer will consider how issues of gender, race or ethnicity, and other social factors relate to the story.

Clients tend to talk about what you will listen to Your task is to listen to the client, but what you listen for may affect the way the client talks about her or his issues. For example, if you pay attention to dreams and past experience, your client will tend to talk about these topics. If, like the Rogerian, you focus on self-talk, the client will talk about her- or himself. If you pay attention to behaviors and thoughts, expect the client to talk in a similar vein and if you focus on cultural or gender issues, these will be prominent.

This means you have considerable power as a listener. It is important that you notice your style and patterns of listening, for what you listen to is an indication of your inherent (and often unconscious) theory of helping. Given these facts, it is important that you give special attention to examining yourself via audiotape and videotape and seek consultation and supervision throughout your career as a helper. It is possible for the most experienced of professionals to find that they are missing important data because of blind spots in listening.

Action skills vary among theoretical orientations The cognitive–behavioral helper will tend to do more structuring and be willing to provide advice and information, whereas the Rogerian person-centered theorist often avoids these skills. Both may provide feedback to the client in the process of growth. The psychodynamically oriented professional will be interested in interpreting and reframing client experience from the perspective of psychodynamic theory. The feminist helper will often orient women clients to issues of societal sexism.

In short, there are important commonalities in helping skills among varying theories, but at the same time, there are important differences. Always recall that the basic skills in this book influence the way your client will think about and interpret his or her problems and issues. Seek consultation and supervision throughout your career, no matter how experienced and expert you become.

Using This Book

This programmed text is based on the single-skills concept. The text presents both positive and negative instances of interviewing behavior. You will be presented with some information—an interviewing problem or issue—and be asked to respond to a question. You will then be directed to the next frame, in which your answer to the question is evaluated. If your answer is correct, we will explain why it is correct and ask you to continue. If your answer is incorrect, we will explain why it is wrong and ask you to return to the original frame to select another answer. It is more effective to not move ahead until you have mastered each step of the program.

The technique used in this book was developed by N. A. Crowder, who demonstrated that mistakes can be considered opportunities to receive additional clarification on critical issues. Research has demonstrated that the reward of accomplishment is important in programmed-learning formats. In the past decade much of the research on programmed learning has informed the development of computer-assisted learning programs.

Using the materials in this book, you can learn interviewing skills that are useful in many settings. But again, unless you systematically practice and use these skills, the effort of learning them will be wasted. Cognitive learning or merely reading the material is not enough: you must integrate these skills into your style. The skills presented here are the basis for establishing constructive interpersonal relationships as well as for sharing, information gathering, self-exploration, and positive personal change.

Let us now try some more programmed frames and examine the system further.

Research Box

Microtraining's systematic paradigm is the closely analyzed and tested training program in interviewing education. Since the original 1968 study on attending behavior, over 450 studies on microskills have been completed (see Daniels, 2003; Ivey & Ivey, 2003). A summary of the research reveals the following:

1. *The method of training has construct and predictive validity.* The skills can be taught with quick, observable impact on trainees and their clients.
2. *Clients respond better and more positively to counselors who attend to them.* They are more satisfied and more likely to return for further interviews if you engage in listening and attending skills.
3. *Different theories have varying patterns of skill usage.* For example, you will find that Rogerian person-centered helpers tend to use more reflective skills, whereas

cognitive–behavioral counselors may be expected to use more action skills.
4. *Gender and multicultural differences exist in skill usage.* Women tend to use more reflective listening skills, whereas men tend to use more questions. African Americans can be expected to focus more on contextual–environmental issues, whereas European North Americans tend to emphasize individual issues more frequently.
5. *The programmed-learning format of* Essential Interviewing *is an effective way to teach skills.* A student who works through the programmed frames of this book may be expected to demonstrate understanding and mastery of skills.
6. *If practiced, skills will be maintained over time.* However, if you do not use the skills beyond this book, they can be easily lost over time. Data are clear that the skills make a difference in interviewing practice, but only if you actually use them in real-life situations.

1.9

Imagine you have now spent a half hour with Susan. During this time, she has been tearful several times and she has mentioned that Chris has been assaultive on many occasions—the bread knife was only a minor example of what she has faced. Susan tells you that their relationship is "not too bad" except when Chris gets depressed and starts drinking. You also have learned that she has left Chris once before, but soon returned when she learned she had minimal job skills and couldn't find employment.

As the interview has progressed, you have increasingly noted her fears and lack of self-esteem, a frequent pattern for women who have suffered assault.

SUSAN: (tearfully) Chris was drinking when I left. He had fallen asleep, and so I felt it was safe to come to you now. When he wakes, he may be fine, but . . . I don't know.

How would you respond?

INTERVIEWER: Susan, it sounds like you're terribly confused. You're not sure which way to turn. (Go to 1.10)

INTERVIEWER: Susan, let's explore just a bit more. We've talked about the positives and negatives of staying and leaving. Which seems to make most sense to you now? (Go to 1.11)

INTERVIEWER: Susan, you've talked about being hit, being burnt by cigarettes—and you've told me that Chris has even threatened your life once or twice when he was especially drunk. He refuses to seek help for his alcoholism. We both know that he isn't likely to change in the immediate future. Your safety is a prime issue to me. It doesn't sound safe for you to return home right now. I'd like to put you in touch with a shelter for tonight. Tomorrow morning, I'll be there with you and we can talk further. (Go to 1.12)

1.10 Your answer: *Susan, it sounds like you're terribly confused. You're not sure which way to turn.*
This is a good reflective response, but it is inappropriate at this stage of the interview. Early in the interview, such a response is important, as it helps bring out data. However, at this point, Susan's feelings have been explored and clarified. What was once a good response now becomes potential "wheel spinning" and involves needless repetition.

This response is somewhat similar in structure to the reflective response of 1.2. Turn back to 1.2 at the beginning of the chapter and note the similarity. The example illustrates the importance of noting where the client stands in exploring the situation and of changing your response to meet differing client needs.

1.11 Your answer: *Susan, let's explore just a bit more. We've talked about the positives and negatives of staying and leaving. Which seems to make most sense to you now?*
Once again, this is an effective response if it is timed to the immediate needs of the client. However, the interviewer has already explored the positives and negatives of staying and leaving and has determined that immediate safety is an issue. It is time for a more active approach.

1.12 Your answer: *Susan, you've talked about being hit, being burnt by cigarettes—and you've told me that Chris has even threatened your life once or twice when he was especially drunk. He refuses to seek help for his alcoholism. We both know that he isn't likely to change in the immediate future. Your safety is a prime issue to me. It doesn't sound safe for you to return home right now. I'd like to put you in touch with a shelter for tonight. Tomorrow morning, I'll be there with you and we can talk further.*
Correct. This response is timed to the current needs of the client. The crisis has been explored, the positives and negatives of leaving have been examined, and the counselor has determined that there is an immediate need to protect the safety of the client.

Evaluating "Correctness" of Responses

The theoretically "correct" answer, however, remains problematic. Many women still return to unsafe situations because the alternative of being on their own is frightening. If Susan decides to return to Chris, there is nothing that you can do at the moment. You have honestly shared your beliefs, and the client has decided to do something else.

On the other hand, if Susan does go to the shelter, you have the responsibility to help her plan for the future. This undoubtedly involves providing her with personal and vocational counseling and helping her find financial assistance and develop a support network. Individual counseling alone cannot solve all these problems. As an effective interviewer, you will want to know how to use community resources such as support groups, church or other crisis-support teams, and community or state agencies.

Even your best intervention with some clients will not be enough. At times, you will want to involve others with the client in a comprehensive treatment plan. It should be clear that only listening to a client may not be sufficient. At times, there is a need for action.

Fortunately, not every case you work with will be as complex as the case of Susan. But if you become a professional helper, you will encounter many cases as fully complex and difficult as this one.

Essential Interviewing does not purport to supply you with all the answers of interviewing, counseling, and therapy. What it can do is provide some basic guidelines of listening and action skills, which you ultimately will have to integrate in your own fashion.

The critical importance of timing responses Particularly important in the last set of programmed responses is the issue of timing. All three listed responses are potentially effective, but unless they are used in a timely manner with the immediate needs of the client, your intervention will fail. Thus, this book asks you to contemplate two equally important tasks:

1. *Learning the basic foundational skills of helping* Through this task, you will develop a flexible skill repertoire that will enable you to have many different responses available for your varying clientele. If your first response to a client doesn't work, you will have several additional alternatives ready to use.
2. *Integrating the skills in a timely and natural fashion* What is an effective helping lead with one client may be harmful to that same client at a later time. As you work through this book, think about how you would naturally integrate and time the suggested skills. If you use these skills in a wooden and mechanical way, you will tend to be ineffective—you might even be dangerous. If you focus on a client's special needs and provide your expertise to her or him in a timely, natural fashion, you are well on your way to using *Essential Interviewing* as part of an effective treatment plan.

An Overview of the Skills Covered in This Book

This book provides the fundamental skills essential for effective interviewing. We have organized the skills into the four sections that follow. Whereas skills and strategies are essential, it is you who will do the personalizing and integrating and ultimately "put it all together."

The following summary should help you see where you are headed and why.

Developing Listening Skills to Draw Out Client Experiences, Issues, and Problems

1. *Attending behavior* (Chapter 2) sharpens your general listening skills as well as your attending skills of eye contact and appropriate verbal and nonverbal following.
2. *Effective questioning* (Chapter 3) teaches you to ask open and closed questions and to make minimal encouragements. You will also learn their usefulness in helping clients express themselves fully.
3. *Reflecting content* (Chapter 4). Effective listening requires more than attending, questioning, and tuning in to emotions. It also requires the ability to *hear* and reflect clearly the verbal content of a client's statements.
4. *Reflecting feeling* (Chapter 5) teaches the accurate identification and reflection of the client's emotions and gives you the opportunity to learn and practice this skill.

Personalizing the Interview

5. *Developing an individual style* (Chapter 6) reviews the preceding skills in an interview that allows you to see how well you have mastered the material thus far. Integrating the skills and developing a personal style are considered.

Helping Clients Understand and Expand Their Experiences

6. *Communicating feeling and immediacy* (Chapter 7). Sometimes it is important to share your own *immediate* feelings in the interview with the client. This chapter presents examples of appropriate and inappropriate communications of feeling and immediacy.
7. *Confronting* (Chapter 8). Specifically, confrontation involves the identification of discrepancies or inconsistencies in a client's behavior. Before you confront a client, however, you must be able to identify inconsistencies and state them clearly without value judgments.
8. *Self-disclosing* (Chapter 9). At times, it is appropriate for an interviewer to self-disclose. However, this must be done without taking over an interview or leaving the client little room for self-exploration.
9. *Information giving* (Chapter 10). All interviewers are required to provide information to clients from time to time. This may take the form of an orienting statement, instructions, feedback, a reframe, or an informational statement.

10. *Interpreting* (Chapter 11). This chapter enables you to identify the underlying meaning of the client's narrative and to reconstruct this meaning for the client.

Taking Action

11. *Structuring for information and action* (Chapter 12) helps you conduct interviews so that clients are able to examine issues, consider alternatives, and make their own decisions.
12. *Enlisting cooperation* (Chapter 13). It is normal for clients to experience some degree of discomfort or reluctance during an interview. This chapter assists you in applying basic interviewing and action skills to decrease client discomfort, increase client cooperation, and help clients move through the stages of change.
13. *Putting it all together* (Chapter 14) integrates the preceding skills. The challenge of this book is manifest most completely in this chapter. You will move through an interview, selecting from an array of potential responses. Once you have mastered the material in the preceding chapters, integration of the skills and concepts will come naturally to you. However, mastery is not achieved by reading alone. This book can be successful only when you take the concepts out of each chapter and practice them in interviewing situations.

All these skills and strategies lead to what is termed "intentionality" (see Ivey & Ivey, 2003). Intentionality, for purposes of this book, could be summarized as follows:

1. Having many responses available to facilitate client growth and development. *Essential Interviewing* seeks to provide you with multiple skills for helping your client.
2. Anticipating how clients will respond to your interviews. The intentional interviewer not only selects skills but also knows what is likely to happen when a certain skill is used. For example, if you reflect content expect the client to elaborate on content issues, but if you reflect feeling, anticipate emotional discussion.
3. Being fully aware that clients are unique, and no matter how skilled and thoughtful your skill or intervention, they will often respond in unique and individual ways.
4. Having another skill or strategy ready when the expected does not happen.
5. Modifying all of the above in response to gender and the multiple issues of diversity.

Intentional interviewing depends on you being flexible and ready to flex and "flow" with what happens in the here and now of the interview. We all make mistakes or say the wrong thing at times—it is your recovery skills that count!

Multicultural and Gender Issues in Interviewing

As we have stressed, patterns and styles of communication vary from culture to culture and between men and women in each culture. In European North American culture, the attending and listening patterns vary from other cultural norms. For example, in a classic study, Haase and Tepper (1972) found that interviewers who lean toward a client, maintain steady eye contact, and establish a conversational distance of slightly over an arm's length are considered most effective. But do these findings hold true universally?

Listening skills are foundational and important in all cultures—the microcounseling approach has been translated into at least 13 languages and is used throughout the world. Estimates are that over 500,000 professionals and paraprofessionals have experienced some form of microskills training. Through wide use and cultural feedback, we have learned the following cultural difference in the basic listening skills.

- *Eye contact*   The direct European North American pattern of eye contact is considered intrusive and rude in some cultures. Many Native Americans and Dene, Latinos/Latinas, and people from other countries will prefer less eye contact. In fact, some traditional African Americans tend to use eye-contact patterns that are directly opposite to standard European American usage; they may gaze at you more while talking and avoid direct eye contact while listening.

In addition, many clients who are discussing intensively private issues may feel more comfortable if you avoid a direct gaze. Eye contact is critical, but it must be modified to meet individual and cultural needs.

- *Body language and space* Let us recall that in some cultures people shake their heads up and down to indicate "no," in direct contrast to typical North American styles. Consider this a metaphor; body language meaning varies widely from culture to culture.

An arm's length may be comfortable for many North Americans, regardless of ethnic or racial background. However, those of recent Arab descent may be more comfortable with 6 to 12 inches separating conversation partners. Other cultures (e.g., Australian aborigines) prefer much greater distance, particularly when they do not have trust for the ethnically or racially different helper.

- *Verbal following* This book stresses that you must listen to the client and follow his or her words as directly as you can. This style of listening is, again, very North American. A more subtle style of listening, more indirect in nature, may be appropriate at times for a Chinese Canadian or Japanese American.

- *Listening first, then acting* This maxim is basic to the field of counseling and therapy. With all clients, we need to listen to their narrative before coming to a conclusion and acting. In truth, some clients want at least a sign of direct helper action before they trust you enough to share their deeper issues. Thus, listening skills remain vital, but the *timing* of listening and acting may vary. With some clients, you need to act fairly soon or you will lose them.

All this is complicated by the fact that European American white males tend to ask more questions than females. Females tend to use more reflective responses. Some Native Americans and African Americans find advice giving and direct support more essential to establishing relationships.

Consider the photos. Each person you work with is unique, yet she or he is also representative of gender and multicultural experiences. There indeed is a modal type of effective interviewing and counseling in North American and European cultures. The ideas of listening and acting are universal. Yet, it is important for you to recall that the basic framework presented here needs to be modified by gender, culture or race, sexual preference, physical issues that the client faces, age of the client, and a multitude of other factors.

As indicated in the preface (page 8), extensive consideration was given to the method of incorporating the material on culture and gender into each chapter. Due to the introductory nature of this book and the work of the first author in the field of cultural psychology (Kazarian & Evans, 1998, 2001), it was decided that the first priority of this book was the mastery of the single skills in their generic form with supplementary material on their cultural applicability provided at the end of each chapter. While the literature provides some information on the use of the skills with some cultures, this is primarily descriptive and there is as yet little empirical research on the myriad of cultures which interviewers might face. The descriptive material on the single skills' cultural applicability will be found in the "Cultural Considerations" section at the end of each chapter.

Learning how gender and cultural differences play themselves out in the diverse cultures of the United States, Canada, and other countries will be a significant part of your lifelong learning as an interviewer. Following are some important sources for further exploration of these issues.

Hall, E. (1959). *The silent language.* Greenwich, CT: Fawcett. *The book that helped start a scholarly movement. Hall's work was foundational to anthropology, psychology, and counseling in finally discovering a simple, but profound point—cultures vary in the way they communicate.*

Ivey, A., D'Andrea, M., Ivey, M., & Simek-Morgan, L. (2002). *Counseling and psychotherapy: A multicultural perspective* (5th ed.). Boston: Allyn & Bacon. *Counseling and psychotherapy theory is reviewed through gender and multicultural lenses. Specifics for action with culturally varying populations are suggested.*

Kazarian, S., & Evans, D. (Eds.). (1998). *Cultural clinical psychology: Theory, research, and practice.* New York: Oxford University Press. *A model of cultural clinical psychology is developed and its application in the practice of clinical psychology is reviewed.*

Kazarian, S., & Evans, D. (Eds.). (2001). *Handbook of cultural health psychology.* San Diego, CA: Academic Press. *Theoretical considerations of health psychology from a cultural perspective are discussed. Specific health concerns such as cancer and pain are considered from a cultural perspective. Health issues of specific cultural groups are reviewed.*

McGoldrick, M., Giordano, J., & Pearce, J. (1996). *Ethnicity and family therapy* (2nd ed.). New York: Guilford Press. *Culture is primarily learned in our family of origin. This book provides stimulating chapters on many of the most common family structures in North America. It includes chapters on Native American, Mexican, Cuban, French Canadian, German, Irish, and many other families. A revision is currently near completion.*

Ponterotto, J., Casas, J. M., Suzuki, L., & Alexander, C. (2001). *Handbook of multicultural counseling* (2nd ed.) Beverly Hills, CA: Sage. *A comprehensive outline of multiple issues and possibilities in the field. This book covers a wide range of issues from theory to practice, including cultural identity theories, the role of traditional healers, and specific actions that may be taken in the interview.*

Sue, D. W., & Sue, D. (2003). *Counseling the culturally different* (4th ed.). New York: Wiley. *A classic of the helping field, this book outlines many important concepts and details that illustrate how to modify interviewing style to meet the needs of culturally diverse populations.*

Ethical Competence and Interviewing

Interviewing practice rests on a foundation of ethics and multicultural awareness. The American Counseling Association (1995) includes the following in its ethics code preamble:

The American Counseling Association is an educational, scientific, and professional organization whose members are dedicated to the enhancement of human development throughout the life-span. Association members recognize diversity in our society and embrace a cross-cultural approach in support of the worth, dignity, potential, and uniqueness of each individual.

Virtually all professional associations include diversity and ethical issues related to interviewing in their ethical codes and we urge you to examine these codes.[1] Web sites for the codes of a number of these professional bodies are listed at the end of this chapter.

Ivey and Ivey (2003) suggest five key areas of ethical competence that may be useful as you work with this text:

1. Competence: Very few of us are able to work with all clients that come to us. Do you have adequate referral sources and knowledge? Do you know your boundaries of competence? Are you aware of your biases?
2. Informed consent: When working with clients, they should be fully aware of what is going to happen and, usually, written consent is necessary. We suggest written consent even for practice sessions as this will help introduce you to professional ethical practice at an early stage.
3. Confidentiality: Obviously it is necessary to retain client confidences, both as a professional in practice and as a student, but you do not have legal confidentiality. If a client proves dangerous to self or others, you usually have a responsibility to inform others. You will want to study the ethical codes of your profession and familiarize yourself with the policies and procedures of your work setting with respect to confidentiality.
4. Power. The helper always has a certain level of power that the helpee or client does not have. Use that power carefully and respectfully. Dual relationships in which you have more than one form of contact with a client can be a problem. For example, you may conduct a practice session with a classmate who is also a friend—you are functioning daily, as friend and interviewer. Or, as a small-town counselor or social worker, you may counsel a client who is also on a church committee with you. These situations bring about special ethical issues.
5. Social justice. The social work and human service professions have given special attention to preventing social problems as well as remediating them through counseling. They argue that the helper has a special responsibility in the community to fight issues of social injustice. The National Association of Social Workers has a specific ethical statement that comments in part, "Social workers pursue social change, particularly with and on behalf of vulnerable and oppressed individuals and groups of people."

Needless to say, this very brief summary cannot provide adequate coverage of this important set of issues underlying interviewing practice. We urge you

---

[1]As a beginning point for ethical code searches, consider the American Association for Marriage and Family Therapy, American Bar Association, American Counseling Association, American Nurses Association, American Psychological Association, Australian Psychological Association, British Association for Counselling, Canadian Counselling Association, Canadian Psychological Association, National Association of Human Service Education, National Association of Social Workers, and New Zealand Association of Counselors.

to visit several professional association Web sites, discuss this matter with your student colleagues and professors, and work to integrate ethical issues throughout your experience with this text.

References

American Counseling Association. (1995). *Code of ethics and standards of practice.* Alexandria, VA: Author.

Blocksma, D., & Porter, E. (1947). A short-term training program in client-centered counseling. *Journal of Consulting Psychology, 11,* 55–60.

Carkhuff, R. (2000). *The art of helping in the 21st century.* Amherst, MA: HRD Press.

Daniels, T. (2003). A review of research on microcounseling: 1967–present. In A. Ivey & M. Ivey, *Intentional interviewing and counseling: Your interactive resource* [CD-ROM]. Pacific Grove, CA: Brooks/Cole.

Egan, G. (2002). *The skilled helper.* (6th ed.). Pacific Grove, CA: Brooks/Cole.

Haase, R., & Tepper, D. (1972). Nonverbal components of empathic communication. *Journal of Counseling Psychology, 19,* 417–424.

Hearn, M. (1976). *Three models of training counsellors: A comparative study.* London: University of Western Ontario.

Hill, C., & O'Brien, K. (1999). *Helping skills.* Washington, DC: American Psychological Association.

Ivey, A. (1971). *Microcounseling: Innovations in interviewing training.* Springfield, IL: Charles C. Thomas.

Ivey, A., D'Andrea, M., Ivey, M., & Simek-Morgan, L. (2002). *Theories of counseling and psychotherapy: A multicultural perspective* (5th ed.). Boston: Allyn & Bacon.

Ivey, A., & Ivey, M. (2003). *Intentional interviewing and counseling: Facilitating client development* (5th ed.). Pacific Grove, CA: Brooks/Cole.

Ivey, A., Normington, C., Miller, C., Morrill, W., & Haase, R. (1968). Microcounseling and attending behavior [Monograph]. *Journal of Counseling Psychology, 15,* 1–12.

Ivey, A., Pedersen, P., & Ivey, M. (2001). *Intentional group counseling: A microskills approach.* Pacific Grove, CA: Brooks/Cole.

Kazarian, S., & Evans, D. (Eds.). (1998). *Cultural clinical psychology: Theory, research, and practice.* New York: Oxford University Press.

Kazarian, S., & Evans, D. (Eds.). (2001). *Handbook of cultural health psychology.* San Diego, CA: Academic Press.

Kelly, G. (1955). *The psychology of personal constructs* (Vols. 1–2). New York: Norton.

Monk, G., Winslade, J., Crocket, K., & Epston, D. (1997). *Narrative therapy in practice.* San Francisco: Jossey-Bass.

Rogers, C. (1977). *On personal power: Inner strength and its revolutionary impact.* New York: Delacourt Press.

Rogers, C. R. (1942). *Counseling and psychotherapy.* Boston: Houghton Mifflin.

Rogers, C. R. (1957). Training individuals in the therapeutic process. In C. Strother (Ed.), *Psychology and mental health* (pp. 76–92). Washington, DC: American Psychological Association.

White, M., & Epston, D. (1992). *Narrative means to therapeutic ends.* New York: Norton.

Zimmer, J. M., & Park, P. (1967). Factor analysis of counselor communications. *Journal of Counseling Psychology, 14,* 198–203.

InfoTrac® College Edition Keyword Search Terms

Competence AND Counseling
Confidentiality
Counselor Skills
Effective Interviewing
Ethics AND Counsel*
Informed Consent
Listening Skills
Multiculturalism
Programmed Instruction
Psychology Ethics
Social Work Ethics

Codes of Ethics on the Internet

American Association for Marriage and Family Therapy (AAMFT): *http://www.aamft.org/resources/LRMPlan/Ethics/ethicscode2001.htm*

American Counseling Association (ACA): *http://www.counseling.org/resources/ethics.htm*

American Nurses Association (ANA): *http://nursingworld.org/ethics/code/ethicscode150.htm*

American Psychological Association (APA): *http://www.apa.org/ethics*

Canadian Association of Social Workers (CASW): *http://www.casw-acts.ca/English/Library/CodeOfEthicsTofC.htm*

Canadian Counseling Association (CCA): *http://www.ccacc.ca/CCA_Code.pdf*

Chapter Two

Attending Behavior

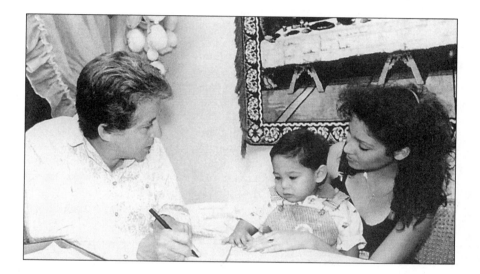

*T*he material in this chapter is intended to help you master the skill of attending. After completing this chapter, you should be able to:

1. Maintain appropriate eye contact during an interview.
2. Maintain a natural, relaxed, and attentive posture during an interview.
3. Demonstrate, by means of your verbal responses, that you are attending to what a client is communicating.
4. Manage appropriate silences during an interview.

The word *attending* is used here to identify the basic skills essential to successful interviewing. These skills must be mastered before other skills can be learned and used effectively. Attending behavior involves both the way you orient yourself physically toward a client, and the manner in which you respond to what the client is communicating. Three important component skills are associated with attending: eye contact, nonverbal behavior, and listening or verbal following. We will consider each of these component skills and then examine the role of silence in interviewing.

Attending behavior facilitates client communication by helping a client communicate in a free and open manner, and keeping interviewer comments to a minimum. On occasion, nonattending behavior can be used purposefully to help clients talk less about destructive or nonproductive topics.

As pointed out in Chapter 1, cultures and races differ in the use of eye contact, other nonverbal behavior, and silence. Because of the diversity of these differences, it is impossible to consider all of them in this chapter. When you deal with people whose cultural background differs greatly from your own, amend your behavior accordingly. Allow your nonverbal behavior to convey your interest and attention appropriately.

The program in this chapter is based on an interview between a nurse (male or female) and a client in a rehabilitation unit. The client is a married female patient who has recovered from major surgery. However, she has shown reluctance to leave the hospital and resume her normal daily activities. Go to 2.1 to begin this program.

2.1

Very few people in our society are effective listeners; most of us find it difficult to focus our attention on others and their comments. One of the component skills of effective listening is appropriate eye contact.

INTERVIEWER: Could you tell me what is making it difficult for you to return to your home?

CLIENT: Well, it's difficult to know where to start. I guess things just got to be too much for me. I seem to be tense all the time.

Choose the most appropriate response.

INTERVIEWER: (Turn to the side and take notes.) (Go to 2.2)

INTERVIEWER: (Because you might upset the client, avoid looking directly at her.) (Go to 2.3)

INTERVIEWER: (Look directly at the client and, with your eyes, encourage her to continue talking.) (Go to 2.4)

2.2   Your answer: *(Turn to the side and take notes.)*

Unless it is absolutely necessary, do not take notes during an interview; it's best to make them after the interview, when you are more aware of key issues. Taking notes during an interview can distract and upset a client. To indicate that you are interested in what a client is saying, maintain appropriate eye contact. Return to 2.1 and choose the response that indicates you are following the client.

2.3	**Your answer:**	*(Because you might upset the client, avoid looking directly at her.)*

You're likely to upset the client by not looking at her. Direct eye contact tells her you're interested in what she is saying. Return to 2.1 and try again.

2.4	**Your answer:**	*(Look directly at the client and, with your eyes, encourage her to continue talking.)*

Correct. When you talk to someone, look at her or him. Do not stare; simply be aware that you are talking to another person. Go to 2.5.

2.5

The amount of eye contact you maintain is an obvious indication of your interest or lack of interest in a client. Therefore, be aware of "eye-contact breaks." Frequent eye-contact breaks can indicate discomfort or a lack of interest in what a client is discussing.

INTERVIEWER: Could you tell me about it?

CLIENT: I was afraid to go outside the house. Just thinking about going out into the yard made me tense. I just couldn't do it.

Choose the most appropriate response.

INTERVIEWER: (Maintain eye contact and smile or nod as the client talks.) (Go to 2.6)

INTERVIEWER: (Look out the window while you listen to the client.) (Go to 2.7)

INTERVIEWER: (Look around the room.) (Go to 2.8)

2.6	**Your answer:**	*(Maintain eye contact and smile or nod as the client talks.)*

Correct. By maintaining eye contact with the client, you indicate that you are focusing on her and following what she is telling you. Go to 2.9.

2.7	**Your answer:**	*(Look out the window while you listen to the client.)*

Although you are listening to the client, you are not maintaining eye contact. The client may assume you aren't listening. Direct eye contact indicates that you are interested in what she is saying. Return to 2.5 and try again.

2.8	**Your answer:**	*(Look around the room.)*

This behavior indicates that you aren't interested in what your client is saying. The first step in helping a client is to follow carefully what he or she is saying by listening and by maintaining eye contact. Return to 2.5 and try again.

2.9

Develop a natural, culturally appropriate style of eye contact with a client. Don't stare or avoid a client's gaze. Brief eye-contact breaks may occur when either the interviewer or client is thinking and when natural breaks in the discussion occur. Eye contact can also provide feedback, and signal understanding and turn taking.

INTERVIEWER: Uhm-hmm.

CLIENT: It's not that I don't want to go out. I do. But when I try to leave the house, I feel anxious and can't do it.

Choose the most appropriate response.

INTERVIEWER:	(Avoid the client's anxious gaze.) (Go to 2.10)	
INTERVIEWER:	(Maintain varied, culturally appropriate, natural eye contact to show your interest.) (Go to 2.11)	
INTERVIEWER:	(Keep your eyes on the client's eyes at all times to show your interest.) (Go to 2.12)	

2.10   Your answer: *(Avoid the client's anxious gaze.)*
An interviewer should respond to a client's fears and help allay them. If eye contact is culturally appropriate and you avoid a client's gaze, you'll probably add to his or her fears. Return to 2.9 and try again.

2.11   Your answer: *(Maintain varied, culturally appropriate, natural eye contact to show your interest.)*
Correct. By maintaining varied, culturally appropriate, natural eye contact, you indicate your interest in and understanding of what a client is saying. Go to 2.13.

2.12   Your answer: *(Keep your eyes on the client's eyes at all times to show your interest.)*
An unwavering stare on the part of an interviewer can be very disconcerting to a client. The use of natural eye contact indicates that you are focusing on and are interested in what a client is saying. Return to 2.9 and try again.

2.13   *Interviewing is not merely a verbal relationship. Much of our communication is nonverbal. Clients know you are interested when you assume a relaxed, professional body position, use appropriate facial expressions, and engage in meaningful gestures. The prudent use of head nodding can signify to clients that you are listening.*

INTERVIEWER:	You try to go out, but you can't.
CLIENT:	Yes. I want to be able to go to work again and visit people, but just can't do it.

Choose the most appropriate response.

INTERVIEWER:	(Sit in a relaxed, professional manner.) (Go to 2.14)
INTERVIEWER:	(Sit in an erect, rigid manner.) (Go to 2.15)
INTERVIEWER:	(Sit back, relax, and put your feet on the desk.) (Go to 2.16)

2.14   Your answer: *(Sit in a relaxed, professional manner.)*
Correct. This is an appropriate way to sit during an interview. Remember that your use of appropriate posture, understanding facial expressions, and encouraging gestures such as head nodding indicate that you are interested in what the client has to say. Go to 2.17.

2.15   Your answer: *(Sit in an erect, rigid manner.)*
If you sit rigidly, the client will find it difficult to relax. If you assume a natural, relaxed position, both you and the client will feel free to relax. Return to 2.13 and try again.

2.16 **Your answer:** *(Sit back, relax, and put your feet on the desk.)*
If you seem to be too relaxed, the client might feel she is boring you, and she might hesitate to discuss her difficulties with you. Your posture should indicate that you are at ease in her presence and that you are focusing on her problems. Return to 2.13 and try again.

2.17 *As an interviewer, be aware of your tone of voice and your rate of speech. Some vocal behaviors communicate caring and involvement; others tend to alienate clients.*

CLIENT: It's so much worse when I try to go out. I get tense and can't do it, and then I get very depressed about it.

Choose the most appropriate response.

INTERVIEWER: (in a rapid, abrupt voice) Tell me how it all started. (Go to 2.18)

INTERVIEWER: (in a warm, expressive voice) Tell me how it all started. (Go to 2.19)

INTERVIEWER: (in a bored voice) Tell me how it all started. (Go to 2.20)

2.18 **Your answer:** *(in a rapid, abrupt voice) Tell me how it all started.*
Your rapid, abrupt voice may give the client the impression that she is wasting your time and pressure her to proceed too quickly. Many clients are unable to proceed when they feel pressured. Your task is to help the client relate to you. Return to 2.17 and try again.

2.19 **Your answer:** *(in a warm, expressive voice) Tell me how it all started.*
Correct. Always use a warm, expressive voice when you conduct an interview, helping the client relax and concentrate on the problem under discussion. Go to 2.21.

2.20 **Your answer:** *(in a bored voice) Tell me how it all started.*
If the client realizes you are bored, she is unlikely to reveal important information about herself. Your task is to communicate your interest in her. Return to 2.17 and try again.

2.21 *Interviewers often attempt to solve problems too quickly by offering sympathy and solutions before a client has had an opportunity to tell his or her story. You'll accomplish more with a client if you are a good listener and refrain from giving advice prematurely. It is essential that you relax and take time to listen to the client carefully.*

CLIENT: I fainted in the bank about 4 years ago, and when I came to, a lot of people were standing over me.

Choose the most appropriate response.

INTERVIEWER: But that was 4 years ago. It shouldn't bother you now. (Go to 2.22)

INTERVIEWER: You should try to go back to the bank every day until your confidence is reestablished. (Go to 2.23)

INTERVIEWER: You must have felt quite embarrassed when that happened. (Go to 2.24)

2.22   Your answer: *But that was 4 years ago. It shouldn't bother you now.*
This response tends to play down the importance of the client's problem when she has just begun to tell you about it. Effective attending behavior involves focusing on what a client says so she will go on talking. Try to refrain from giving advice. Return to 2.21 and try again.

2.23   Your answer: *You should try to go back to the bank every day until your confidence is reestablished.*
To listen effectively, an interviewer must do more than provide premature answers and sympathy. Your task is to listen carefully, respond to the client, and focus on the topic being discussed. Return to 2.21 and try again.

2.24   Your answer: *You must have felt quite embarrassed when that happened.*
Correct. This response reflects how the client felt in the situation and indicates that you are following her. It also encourages her to continue telling you about her experiences. Go to 2.25.

2.25   *An important interviewing skill is verbal following. As an interviewer, focus on what has already been said rather than introduce a new topic. This isn't easy, but you will be well on the way to becoming a skillful interviewer when you can stay on topic.*

CLIENT: Yes. And I found it difficult the next time. In fact, I left before I had deposited my money. There were so many people . . .

Choose the most appropriate response.

INTERVIEWER: There were so many people . . . (Go to 2.26)

INTERVIEWER: Which bank was it? (Go to 2.27)

INTERVIEWER: What did your husband think? (Go to 2.28)

2.26   Your answer: *There were so many people . . .*
Correct. This response indicates that you were listening to what the client was saying and are interested in hearing more. Go to 2.29.

2.27   Your answer: *Which bank was it?*
The name of the bank isn't really important information. The client has referred to her next visit to the bank, and you have indicated your lack of attention to what she is saying by changing the topic. As an interviewer, you must listen carefully to the client and stay on topic. Return to 2.25 and try again.

2.28   Your answer: *What did your husband think?*
You have introduced a new topic—the client's husband. This information may be of interest to you, but it is irrelevant to the client at this point in the inter-

view. Your task is to focus your attention on what the client is saying, and your response should reflect this attention. Return to 2.25 and try again.

2.29

Remember to stay on topic. Don't jump from topic to topic or interrupt or distract the client. Take your cues from what the client has said and don't let your mind wander.

CLIENT: Yes. All the people bothered me. Then I began to get upset in other places where there were lots of people.

Choose the most appropriate response.

INTERVIEWER: Excuse me for a moment. What exactly were you doing in the bank? (Go to 2.30)

INTERVIEWER: Could you tell me more about that? (Go to 2.31)

INTERVIEWER: (Interrupt her after she has said, "All the people bothered me.") How many people were there? (Go to 2.32)

2.30 Your answer: *Excuse me for a moment. What exactly were you doing in the bank?*
You have interrupted the client, and as a result, you may find it difficult to get back on topic. The client may not be inclined to share important information with someone whose thoughts were obviously elsewhere. Your task is to follow intently what the client is saying. Return to 2.29 and try again.

2.31 Your answer: *Could you tell me more about that?*
Correct. This question allows the client to discuss her problems further. When you respond appropriately, the client will go on to give you more information. Go to 2.33.

2.32 Your answer: *(Interrupt her after she has said, "All the people bothered me.") How many people were there?*
This interruption prevents the client from presenting her problem in the manner that suits her, and it indicates that you think what she is relating to you is unimportant. Refrain from interrupting the client. It is important that, as an interviewer, you attend to *all* information the client communicates to you. Return to 2.29 and try again.

2.33

As an interviewer, you should attend to what the client has said and direct it back to her, seeking to understand her situation.

CLIENT: Well, it soon became unbearable for me to go into stores or to get my hair cut—to do all the things I should do.

Choose the most appropriate response.

INTERVIEWER: Who cut your hair? (Go to 2.34)

INTERVIEWER: How do you buy new clothes? (Go to 2.35)

INTERVIEWER: All the usual things became unbearable. (Go to 2.36)

2.34	**Your answer:**	*Who cut your hair?*
		Because you have changed topics, the client will probably become confused. Listen carefully to the client and direct what she has said back to her for further clarification. Return to 2.33 and try again.

2.35	**Your answer:**	*How do you buy new clothes?*
		It is unlikely that the answer to this question will assist you in helping the client with her problem. You've changed the topic the client was discussing by pursuing a topic about which you are curious. As an interviewer, pay careful attention to what the client says in order to pick up some aspect of her communication and direct it back to her. Return to 2.33 and try again.

2.36	**Your answer:**	*All the usual things became unbearable.*
		Correct. Your response is on topic and invites the client to tell her story. Go to 2.37.

2.37		*Don't talk too much during the interview. The client has come to you with a problem; she should do most of the talking. There is no need to talk about yourself or to give answers to the client. Your responsibility is to help the client find her own answers.*
	CLIENT:	Yes. When I go out, my stomach gets upset. I get a headache, my throat is tight, and I can't breathe.
		Choose the most appropriate response.
	INTERVIEWER:	How do you think this relates to what happened in the bank? (Go to 2.38)
	INTERVIEWER:	Well, I fainted in public once, too, and I never have those problems. (Go to 2.39)
	INTERVIEWER:	Well, I don't see any reason for you to feel that way. I don't think other people feel that way. (Go to 2.40)

2.38	**Your answer:**	*How do you think this relates to what happened in the bank?*
		Correct. With this response, you are asking the client to explain how her physical reactions in crowded situations relate to her initial fainting spell in the bank. Go to 2.41.

2.39	**Your answer:**	*Well, I fainted in public once, too, and I never have those problems.*
		This information doesn't help the client. She is aware of the fact that she doesn't react the way other people do. Moreover, you've changed the topic and indicated that you are interested in discussing yourself rather than the client. Return to 2.37 and try again.

2.40	**Your answer:**	*Well, I don't see any reason for you to feel that way. I don't think other people feel that way.*
		This response denies the client the right to feel the way she does. Moreover, implying that she should exert more control over her behavior prevents her from clarifying what she is trying to tell you. Return to 2.37 and choose the

response that indicates you are focusing on what the client is saying and are interested in helping her find her own answers.

2.41

An interviewer's comments should help the client clarify the problem. Comments that encourage client dependency or originate from an interviewer's curiosity do not belong in an interview.

CLIENT: Well, I guess when I get into situations in which there are people, I'm always afraid I'll faint. And so I become tense. I'm all right here, though.

Choose the most appropriate response.

INTERVIEWER: Have you told anyone else about this? (Go to 2.42)

INTERVIEWER: Can you clarify for me whether it is the people or the situations that cause you to be afraid? (Go to 2.43)

INTERVIEWER: Well now, we'll have to see what we can do about that. But don't worry anymore. (Go to 2.44)

2.42 **Your answer:** *Have you told anyone else about this?*

This response reflects curiosity rather than listening skills. With this response, you have changed the topic and neglected to help the client clarify her concerns. Return to 2.41 and try again.

2.43 **Your answer:** *Can you clarify for me whether it is the people or the situations that cause you to be afraid?*

Correct. This is an appropriate response because it asks the client to clarify for both of you the cause of her fear of fainting. Go to 2.45.

2.44 **Your answer:** *Well now, we'll have to see what we can do about that. But don't worry anymore.*

By using this patronizing statement, you take on all the responsibility for solving the client's problem. Unfortunately, it is seldom possible to solve a client's problem immediately, and therefore, a well-intentioned response such as this can be misleading. Concentrated listening will enable you to attend to what the client tells you and will help her explore the topic she is pursuing. Return to 2.41 and try again.

2.45

If you find yourself unsure of what to say to the client next, go back and ask a question or make a comment about a relevant topic that was discussed earlier in the immediate or near past.

CLIENT: I don't know. I guess I'm afraid of both the people and the situations. Right now I stay away from both, so I can't really say. I don't know what else to say.

Choose the most appropriate response.

INTERVIEWER: You were telling me earlier about not being able to go out to the bank. What happened next? (Go to 2.46)

INTERVIEWER: From what you've said, I think I know. (Go to 2.47)

INTERVIEWER: (Remain silent and wait for the client to continue.) (Go to 2.48)

2.46	**Your answer:**	*You were telling me earlier about not being able to go out to the bank. What happened next?*

Correct. When you reach a dead end in a discussion, a completely new topic might seem appealing. However, it is better to refer to a topic the client discussed with you in the immediate or near past—something that is of concern to her. This is good attending behavior. Go to 2.49.

2.47	**Your answer:**	*From what you've said, I think I know.*

This response directs the interview away from what the client is discussing. Pause for a moment, think back over what the client has been discussing earlier, and reintroduce a point you would have followed if you'd had the opportunity. Return to 2.45 and try again.

2.48	**Your answer:**	*(Remain silent and wait for the client to continue.)*

The client has told you that she has reached an impasse. Your silence at this point may be appropriate, but because the client is unable to clarify the source of her problem, you are responsible for making the next statement. Pause for a moment, think back over what the client has told you, and then reintroduce a relevant aspect of her difficulties into the discussion. Return to 2.45 and try again.

2.49		*In many cases, a client needs only a little encouragement to continue talking.*
	CLIENT:	Well, as I told you, I stopped going out. Then I found that I couldn't go out socially or visit with friends anymore.
		Choose the most appropriate response.
	INTERVIEWER:	Did your husband complain? (Go to 2.50)
	INTERVIEWER:	It must have been difficult for everyone in your family to make excuses for you all the time. (Go to 2.51)
	INTERVIEWER:	And then . . . (Go to 2.52)

2.50	**Your answer:**	*Did your husband complain?*

You have changed the topic by introducing the client's husband into the discussion. Skillful interviewers stay on topic and allow clients to introduce new topics into the discussion. Return to 2.49 and try again.

2.51	**Your answer:**	*It must have been difficult for everyone in your family to make excuses for you all the time.*

You aren't following what the client is telling you. You have changed the topic and focused on the difficulty she may have caused her family. Your response should focus on the client and her problem and encourage her to continue. Return to 2.49 and try again.

2.52   **Your answer:**   *And then . . .*
Correct. This minimal encouragement invites the client to go on with what she was telling you. Go to 2.53.

2.53

Effective interviewers stop to listen and are occupied with their clients' communication rather than with thoughts of what to say next.

CLIENT: For a while, I couldn't go out socially unless I'd had a few drinks.

Choose the most appropriate response.

INTERVIEWER: Could you tell me more about that? (Go to 2.54)
INTERVIEWER: What did you drink? (Go to 2.55)
INTERVIEWER: Are you an alcoholic? (Go to 2.56)

2.54   **Your answer:**   *Could you tell me more about that?*
Correct. This question allows the client to go on with what she was discussing. You are free to focus on what she is saying rather than introducing a new topic. Go to 2.57.

2.55   **Your answer:**   *What did you drink?*
You don't need this information at this time. Be concerned with the client and her problem, not with areas that are unimportant to what she is telling you. If this information is important to her, she will tell you later in the interview. Return to 2.53 and try again.

2.56   **Your answer:**   *Are you an alcoholic?*
You're overinterpreting what the client has told you. At this point, you don't have sufficient information to ask this question. Moreover, it's possible that the client will be insulted by this question. Stay with what the client is saying. Preoccupation with your own thoughts can result in an ineffective interview. Return to 2.53 and try again.

2.57

As an interviewer, you must attend not only to the verbal cues given by the client but to the nonverbal cues as well—glances, gestures, bodily reactions, tones of voice, and pauses.

CLIENT: (looking down at the floor) Well, I . . . started having . . . just a few drinks.

Choose the most appropriate response.

INTERVIEWER: I'm here to listen. I can't help you if you won't tell me everything. (Go to 2.58)
INTERVIEWER: Maybe we should talk about something else. (Go to 2.59)
INTERVIEWER: You find it difficult to tell me about it. (Go to 2.60)

2.58 **Your answer:** *I'm here to listen. I can't help you if you won't tell me everything.*
The client is aware of your role and doesn't need to be reminded of your position. You have changed the topic from the client to yourself and your role. Respond instead to the verbal and nonverbal cues given by the client. Return to 2.57 and try again.

2.59 **Your answer:** *Maybe we should talk about something else.*
You have responded to the client's verbal and nonverbal cues in discussing her drinking. Unfortunately, your response enables her to avoid rather than to explore the topic. Return to 2.57 and try again.

2.60 **Your answer:** *You find it difficult to tell me about it.*
Correct. This is a good response because it indicates that you were attending to the client's gesture (looking down) and her pauses in speech. It is obviously difficult for her to talk to you about her drinking. By reflecting how she feels, you give her a chance to consider her feelings before going on with the discussion. Go to 2.61.

2.61 *Another skill in interviewing is the effective use of silence. A good interviewer respects the client's need to be silent. You need not fill a silence unless an impasse has been reached. Sit back and give the client time to think.*

CLIENT: (sits silently for a minute and appears to be thinking)

Choose the most appropriate response.

INTERVIEWER: What's the matter? Cat got your tongue? (Go to 2.62)

INTERVIEWER: How many drinks do you have during an average day? (Go to 2.63)

INTERVIEWER: (sits attentively and waits for the client to continue) (Go to 2.64)

2.62 **Your answer:** *What's the matter? Cat got your tongue?*
You have responded to the client's silence, but not in a way that will make her realize that you understand her difficulty in talking about drinking. Allow the client to be silent. Clients often need time to work things out before they can continue in an interview. Return to 2.61 and try again.

2.63 **Your answer:** *How many drinks do you have during an average day?*
The client is uncomfortable with what she is about to discuss, and she needs your acceptance and support while she is thinking out her response. Your efforts to pressure her into talking will have little effect if she is not ready to talk. Return to 2.61 and try again.

2.64 **Your answer:** *(sits attentively and waits for the client to continue)*
Correct. Silence is appropriate when the client is thinking and trying to decide how to continue with something that is difficult to discuss. At this point, the client needs your supportive silence. She will continue as soon as she is ready. Go to 2.65.

2.65

As an interviewer, remember that silence can be a positive form of communication. Some clients need time to think, whereas others use silence to signify agreement or as a sign of respect. Your task is to be sensitive to the meaning of the client's silence and react accordingly.

CLIENT: It's hard to talk about it. I haven't told anyone before, but I guess I should. You see, I drink from the time I get up until my husband gets home. (Client falls silent.)

Choose the most appropriate response.

INTERVIEWER: Does your husband drink? (Go to 2.66)

INTERVIEWER: I guess it's not an easy thing to talk about. (Go to 2.67)

INTERVIEWER: Who pays for all the liquor? It must cost a lot of money. (Go to 2.68)

2.66 Your answer: *Does your husband drink?*
You've changed the topic. The client is telling you about her problem at the moment, not about any problem that her husband might have. Your task is to stay on topic and to stay in tune with the client's behavior. Return to 2.65 and try again.

2.67 Your answer: *I guess it's not an easy thing to talk about.*
Correct. Your response reflects the client's obvious difficulty in talking about her drinking and is supportive of her attempt to discuss her problem. Moreover, this response indicates that you are in tune with her feelings and understand what her silences mean. Go to 2.69.

2.68 Your answer: *Who pays for all the liquor? It must cost a lot of money.*
This information may be important later in the interview, but at this point, you should focus on the client's behavior and help her explore her difficulties. Return to 2.65 and try again.

2.69

You have now completed the teaching frames on attending behavior, the skills that are basic to all successful interviewing. Remember that, as an interviewer, your job is to attend to the other person's story. You do this by maintaining appropriate eye contact, expressive vocal tone, an easy, attentive posture, and by verbally following the client's communications. As you become more skilled in attending behavior you will find that when client talk is inappropriate, repetitive, or negative, it may be ineffective to use attending behaviors. At such times failure to maintain eye contact, changes in body posture, and deliberate shifts to more positive topics can facilitate the interview.

2.70

Review Questions

Indicate whether each of the following statements is true or false.

1. Good interviewers maintain intent eye-to-eye contact at all times.

2. It is most important to offer solutions to a client's problem as soon as possible in the interview.
3. A competent interviewer is comfortable with brief reflective silences.
4. A relaxed, professional, and respectful posture communicates an interviewer's concern for a client.
5. When an interviewer feels temporarily lost during an interview, it is usually a good idea to focus the discussion on something the client has related in the immediate or near past.
6. A competent interviewer is more attentive to verbal cues than to nonverbal cues.

Review Answers

1. *False.* If you answered *true,* go to 2.9 and review.
2. *False.* If you answered *true,* go to 2.21 and review.
3. *True.* If you answered *false,* go to 2.61 and review.
4. *True.* If you answered *false,* go to 2.13 and review.
5. *True.* If you answered *false,* go to 2.45 and review.
6. *False.* If you answered *true,* go to 2.57 and review.

If three or more of your responses were incorrect, return to 2.1 and work through this chapter again.

Points to Remember about Attending Behavior

1. EYE CONTACT:
   a. Suggests that you are attending to what is being communicated.
   b. Should be natural and culturally appropriate.
   c. Is most likely to be modified when a break in discussion occurs or when either party is thinking.
   d. Can signal understanding and provide feedback.
2. BODY POSTURE:
   a. Should be natural, attentive, and relaxed, communicating interest.
   b. Gestures should be easy and natural.
   c. Facial expression should be appropriate to the material under discussion.
3. VERBAL RESPONSES:
   a. Should be made in a warm and expressive tone, made at an appropriate pace, and communicate involvement.
   b. Should follow from the client's comments.
   c. Should not change the topic or interrupt the client.
   d. Should relate to previous aspects of the client's story when the topic being discussed is exhausted.
   e. Should be made with regard to both the verbal (content and tone) and the nonverbal (glances, gestures, and other physical reactions) behavior and culture of the client.

4. SILENCES:
    a. May occur, since clients often need time to think.
    b. Are often a positive form of communication.
    c. Should not become excessive without an interviewer response.

2.1

Activity Unit

Find two others who have also read this chapter. Role play brief interviews (no more than five interviewer responses) alternating the client, interviewer, and observer roles. Record each brief interview on audio- or videotape.

Interviewer role Practice maintaining appropriate eye contact, vocal behavior, attentive posture, and verbal following in response to the client.

Client role Be cooperative and provide sufficient information for the interviewer to practice. You can role play the problem of a friend or family member or enact one of the following roles:

1. A young adult whose parents are separating.
2. A student unsure of what career to pursue.
3. A resident upset about the noise level in the neighborhood.
4. An individual who has just been fired for lateness.

Observer role Operate the equipment and complete the Practice Interview Checklist (see Table 2.1).

Postinterview discussion Using the taped interview and the Practice Interview Checklist data, the participants should discuss the performance of the interviewer. All participants can learn from this discussion, which should be nonjudgmental, focus on positive as well as less effective responses, and stimulate improved performance for all.

2.2

Activity Unit

Using the tape of the interview produced in Activity Unit 2.1 or the tape of another practice interview, enter the original interview responses in the spaces provided. Then formulate a better response to replace each original one in the space provided.

Original response 1 _____

_____

_____

Table 2.1

Practice Interview Checklist

Activity	**Interviewer response 1** C[a]	I[b]	**2** C	I	**3** C	I	**4** C	I	**5** C	I
Eye contact										
Facial expression										
Gestures/posture										
Verbal tone/pace										
Verbal response										

[a] Interviewer response while client (C) is speaking. If the interviewer's activity (eye contact, facial expression, and so on) is appropriate, enter an **A;** if it is inappropriate, enter an **I.**

[b] Interviewer (*I*) response to client's communication. If the interviewer's activity (eye contact, facial expression, verbal response, and so on) is appropriate, enter an **A;** if it is inappropriate, enter an **I.**

Better response ______________________________

____

____

Original response 2 ______________________________

____

____

Better response ______________________________

____

____

Original response 3 ______________________________

____

____

Better response ______________________________

____

____

Original response 4 ______________________________

____

____

Better response _____

_____

_____

Original response 5 _____

_____

_____

Better response _____

_____

_____

Cultural Considerations

1. Politeness is a behavior common to all cultures. However, it is expressed in different ways in different cultures.
2. The degree to which an interviewer's behavior varies from relaxed to formal or from equal to more authoritarian stems from the interviewer's appreciation of the client's culture.
3. Anglo–American interviewers and clients are more comfortable when a greater physical distance exists between them. Some other cultures prefer a closer stance when communicating with each other. Inuit may prefer to sit side by side.
4. Based on the client's cultural affiliation, the interviewer may be required to adjust body language, interviewing position, eye contact, vocal tone, and speech rate. For example, Anglo–Americans engage in more eye contact when listening and less when speaking, whereas the reverse is true for some African Americans. In other cultures eye contact is avoided particularly when discussing serious topics.
5. It is important to determine whether the client's culture permits the interviewer to focus on the client as an individual or on the client as an individual within the context of the family and/or cultural community.
6. In some cultures, clients give brief responses when being interviewed as a sign of respect for the interviewer. This should not be interpreted negatively. In Asian cultures, subtle communication as opposed to direct communication is an art.
7. Silence is viewed in some cultures as an opportunity to reflect on what is being talked about, and the interviewer should accommodate this. English and Arab people use silence to indicate a need for privacy, whereas Russian, French, and Spanish people use silence to signify agreement. In Asian cultures, silence is traditionally a sign of respect for elders. Some Native North Americans use silence to indicate disagreement.

InfoTrac® College Edition Keyword Search Terms

Active Listening
Nonverbal Communication
Silence AND Counseling

Chapter Three

Effective Questioning

This chapter is intended to help you master an effective style of questioning that facilitates communication. Effective questioning requires a knowledge of open (often referred to as open-ended) questions, closed (or closed-ended) questions, and minimal encouragement. After completing this chapter, you should be able to:

1. Formulate and make open questions.
2. Decide whether an open question is appropriate in a given situation.
3. Formulate and make closed questions.
4. Decide whether a closed question is appropriate in a given situation.
5. Identify the forms of questioning that should be avoided.
6. Make minimal encouragements.
7. Decide whether minimal encouragement is appropriate in a given situation.

Interviewers use open questions to encourage clients to talk more freely and openly about their situations, thoughts, and feelings. Open questions can be used to begin an interview, to encourage a client to use more information, and to request specific examples of a client's issues or concerns. Closed questions can be answered with a "yes," a "no," or a simple fact. Closed questions are used to request missing information or a specific fact, ascertain the parameters of a problem, and to manage overtalkative clients who are rambling during the interview. Interviewers use minimal encouragements to indicate to clients that they are attending to what clients are saying, and to encourage clients to continue talking. Minimal encouragements include simple words (*yes, go on,* and so forth), sounds (*uhm-hmm, uh huh,* and so on), and nonverbal behavior (head nods and similar gestures).

In this chapter, open and closed questions are defined, the various forms of open questions are considered, the comparative importance of open and closed questions in interviewing is examined, and situations are studied in which open and closed questions are appropriate. Following a discussion of the forms of questions that should be avoided, the use of minimal encouragements is examined.

The program in this chapter focuses on portions of two interviews conducted by a marriage counselor. The client is a man who has come to discuss his stressful marriage. Go to 3.1 to begin this program.

3.1	*A client comes to an interview to enter into a discussion. The task of the interviewer is to facilitate discussion by providing structure through the use of open questions.*

An open inquiry, often referred to as an open ended question usually requires a more elaborate response than a "yes," "no," or brief factual answer. In contrast, a closed inquiry, or closed ended question, can be answered with "yes," "no," or a simple fact.

CLIENT: I'm having problems with my marriage.

Choose the most appropriate response.

INTERVIEWER: How long have you been married? (Go to 3.2)

INTERVIEWER: Could you tell me more about these problems? (Go to 3.3)

INTERVIEWER: Are you still living with your wife? (Go to 3.4)

3.2   Your answer: *How long have you been married?*
This closed question invites a brief factual answer and interrupts the client's story. An open question provides a less restrictive structure and encourages discussion. Return to 3.1 and try again.

3.3   Your answer: *Could you tell me more about these problems?*
Correct. This open question encourages the client to continue the discussion. Go to 3.5.

3.4	**Your answer:**	*Are you still living with your wife?*

This closed question demands a "yes" or "no" answer and interferes with the client's narrative. An open question is preferable at this point. Return to 3.1 and try again.

3.5

Before you can ask effective questions, you must listen carefully to what the client says. Development of this listening skill depends on your ability to attend to what the client is saying.

CLIENT: My marriage is no good anymore. We're always quarreling. We don't even talk to each other very much anymore. I don't enjoy it.

Choose the most appropriate response.

INTERVIEWER: What do you argue about? (Go to 3.6)

INTERVIEWER: How do you think you could change your behavior? (Go to 3.7)

INTERVIEWER: What was your home like when you were a child? (Go to 3.8)

3.6   **Your answer:**   *What do you argue about?*

Correct. This is an open question that reflects your attention to what the client is discussing. When you ask open questions that are on topic, the client will be inclined to discuss himself and his situation further. Go to 3.9.

3.7   **Your answer:**   *How do you think you could change your behavior?*

This is an open question, but it's premature. You don't know enough about the client to begin seeking solutions to his problem. Moreover, you've shifted the focus from his concerns to yours. Return to 3.5 and try again.

3.8   **Your answer:**   *What was your home like when you were a child?*

This is an open question, but it is off the topic the client is discussing. You'll be of more help to the client if you stay on the topic. Return to 3.5 and try again.

3.9

Four methods are commonly used to introduce an open question. Questions that begin with what frequently elicit factual data.

CLIENT: I guess we fight about a lot of things, but the major thing we fight about is that I'm never home.

Choose the response that enables you to collect factual information about the client's problem.

INTERVIEWER: How do you feel about your situation? (Go to 3.10)

INTERVIEWER: What does your wife do while you're away? (Go to 3.11)

INTERVIEWER: What keeps you away from home? (Go to 3.12)

3.10   Your answer: *How do you feel about your situation?*
This is a good open question, but it will not lead to the factual information you were requested to obtain. Return to 3.9 and try again.

3.11   Your answer: *What does your wife do while you're away?*
This open question is off topic and will not lead to factual information about the client's problem. Return to 3.9 and try again.

3.12   Your answer: *What keeps you away from home?*
Correct. This open question enables you to gather factual information concerning the client's problem. Go to 3.13.

3.13

Open questions that begin with the word *how* encourage a client to give a personal or subjective view of a situation. Therefore, these questions are considered people oriented rather than fact oriented.

CLIENT: My wife is always worried about money, and I have to hold down two jobs to keep up with our bills. That means I work from seven in the morning until ten at night.

Choose the response that enables you to focus the interview on the client's personal view of his situation.

INTERVIEWER: What kind of bills do you have? (Go to 3.14)

INTERVIEWER: What's so terrible about having two jobs? (Go to 3.15)

INTERVIEWER: How do you feel about that? (Go to 3.16)

3.14   Your answer: *What kind of bills do you have?*
This is an open question that will enable you to obtain primarily factual information. Choose an open question that is people oriented—one that focuses on the client and his reaction to his situation. Return to 3.13 and try again.

3.15   Your answer: *What's so terrible about having two jobs?*
This is a leading question: It implies that the client should accept your values. Moreover, this response doesn't encourage the client to consider his personal reaction to his situation. Return to 3.13 and try again.

3.16   Your answer: *How do you feel about that?*
Correct. This open question places the focus of the interview on the client and his feelings. This response is people oriented rather than fact oriented. Go to 3.17.

3.17

Interviewers frequently begin open questions with the words *could, could you,* or *can you. Such questions stimulate detailed responses. This type of open question offers a client flexibility in formulating responses and, therefore, responsibility for contributing to the discussion.*

CLIENT: I'm pretty angry. I work all the time, and then when I am home, I get told about not being home.

Choose the response that offers the client the greatest opportunity to contribute to the interview.

INTERVIEWER: Could you describe one such occasion for me? (Go to 3.18)

INTERVIEWER: What do you do about your anger? (Go to 3.19)

INTERVIEWER: Is that a valid reason to become angry? (Go to 3.20)

3.18   **Your answer:**   *Could you describe one such occasion for me?*
Correct. This open question allows the client to select and describe an incident that is relevant to him. Go to 3.21.

3.19   **Your answer:**   *What do you do about your anger?*
This open question directs the client to focus on a specific topic. Open questions that begin with the words *could* or *can* usually encourage more participation on the part of the client. Return to 3.17 and try again.

3.20   **Your answer:**   *Is that a valid reason to become angry?*
This judgmental and leading question severely limits the client's ability to contribute to the discussion. Your implication that his point of view is incorrect is based on insufficient information. Return to 3.17 and try again.

3.21

The final type of open question discussed in this chapter begins with the word why. Many skilled interviewers tend to avoid this type of question because it can make clients feel defensive.

CLIENT: Well, last night, as soon as I was in the door, my wife started complaining. I was tired and didn't really want to hear her complaints. After all, I have complaints too.

Choose the most appropriate response.

INTERVIEWER: Why did you react that way? (Go to 3.22)

INTERVIEWER: Could you help me to understand what it is about her complaints that upsets you? (Go to 3.23)

INTERVIEWER: Do you know why you didn't want to hear her complaints? (Go to 3.24)

3.22   **Your answer:**   *Why did you react that way?*
This open question may imply that the client should not have reacted the way he did, causing him to feel defensive. Return to 3.21 and try again.

3.23   **Your answer:**   *Could you help me to understand what it is about her complaints that upsets you?*
Correct. This open question permits the client to discuss his problem without becoming defensive. Go to 3.25.

3.24   Your answer: *Do you know why you didn't want to hear her complaints?*
This closed question suggests that you were not attending to the client's last statement. As a result, he may become defensive and withdraw from his interaction with you. Return to 3.21 and try again.

3.25

In skilled interviewing, open questions are usually more important than closed questions. Open questions offer a client opportunities to introduce relevant topics; closed questions limit the client's opportunities to do so.

CLIENT: It's probably her timing. First, it happens night after night; and second, the fact is she never gives me time to relax before she starts.

Choose the most appropriate response.

INTERVIEWER: You don't think she has the right to do that? (Go to 3.26)

INTERVIEWER: Would she stop if you did what she asked? (Go to 3.27)

INTERVIEWER: What do you feel like doing when she starts to complain? (Go to 3.28)

3.26   Your answer: *You don't think she has the right to do that?*
This is a closed question. You've indicated your lack of interest in what the client has said by directing him away from his line of thought to ask him something that is of interest to you. Remember, when you ask an open question, the client has an opportunity to introduce and explore relevant concerns. Return to 3.25 and try again.

3.27   Your answer: *Would she stop if you did what she asked?*
This closed question demonstrates a lack of interest in the client and changes the topic he was pursuing. An open question allows a client to explore his or her concerns with the support of an interviewer. Return to 3.25 and try again.

3.28   Your answer: *What do you feel like doing when she starts to complain?*
Correct. This open question allows the client to introduce and discuss relevant information. Go to 3.29.

3.29

During an interview, questions should elicit information for an interviewer and assist a client in exploring and clarifying his or her concerns. Open questions fulfill both of these functions; closed questions tend to fulfill only the first.

CLIENT: I have two reactions, really. Either I feel like yelling at her and telling her she's not the only one who feels tired and fed up, or I feel like turning around and walking right out of the house.

Choose the most appropriate response.

INTERVIEWER: Could you help me to understand what's going on inside you when you feel this way? (Go to 3.30)

INTERVIEWER: When you yell at her, does she yell back? (Go to 3.31)

INTERVIEWER: Have you ever walked out? (Go to 3.32)

3.30 Your answer: *Could you help me to understand what's going on inside you when you feel this way?*
Correct. This open question is centered on the concerns of the client. This response will enable you to gather information and help the client explore and clarify his problems. Go to 3.33.

3.31 Your answer: *When you yell at her, does she yell back?*
This closed question can be answered by "yes" or "no." As a result, it produces little information and prevents the client from exploring and clarifying his situation. Return to 3.29 and try again.

3.32 Your answer: *Have you ever walked out?*
This closed question provides minimal information. Design your questions to help the client explore and clarify his problems. Return to 3.29 and try again.

3.33

Clients find interviews more enjoyable and satisfying when they're permitted to tell their story in their own way. For this reason, the use of open questions tends to put clients at ease, and the repeated use of closed questions often upsets and irritates them.

CLIENT: My stomach gets all tied up in knots, and my hands clench. I even grit my teeth together.

Choose the most appropriate response.

INTERVIEWER: Have you ever felt so tight inside that you've become physically ill? (Go to 3.34)

INTERVIEWER: Have you ever hit your wife? (Go to 3.35)

INTERVIEWER: I wonder how you understand the way you're reacting? (Go to 3.36)

3.34 Your answer: *Have you ever felt so tight inside that you've become physically ill?*
This closed question might distract and irritate the client. Attempt to acquire this information by using a question that will put the client at ease and permit him to communicate more openly. Return to 3.33 and try again.

3.35 Your answer: *Have you ever hit your wife?*
This response is a closed question. Although it may be important for you to have this information at a later stage, this question interrupts the client's narrative. Interruptions such as this often irritate clients. Return to 3.33 and try again.

3.36 Your answer: *I wonder how you understand the way you're reacting?*
Correct. This open question permits the client to continue discussing his problem with you. As a result, he will probably feel at ease, become more involved, and gain satisfaction from the interview. Go to 3.37.

| **3.37** | | *Open questions are useful in a number of interviewing situations. For example, they can be employed when you want a client to elaborate on a point.* |
| | CLIENT: | I often feel that nobody—not my wife, anyway—really cares about me at all or about how hard it is for me to have to work all the time. They think all I'm good for is bringing home the money. I wish my wife would look at my situation differently. |

Choose the most appropriate response.

INTERVIEWER:	Could you tell me more about that? (Go to 3.38)
INTERVIEWER:	Is the money you bring home enough to justify all the work you do? (Go to 3.39)
INTERVIEWER:	Who are the others who don't really care about you? (Go to 3.40)

3.38   Your answer: *Could you tell me more about that?*
Correct. This open question allows you to help the client elaborate on the point he has just made—the fact that he wants his wife to view his situation differently. Go to 3.41.

3.39   Your answer: *Is the money you bring home enough to justify all the work you do?*
Your response is a closed question that requires only a "yes" or "no" answer. At this point in the interview, an open inquiry will help the client elaborate on his previous comment. Return to 3.37 and try again.

3.40   Your answer: *Who are the others who don't really care about you?*
This closed question will elicit limited factual information. This question won't encourage the client to elaborate on the point he has just made concerning his wife's attitude. Return to 3.37 and try again.

| **3.41** | | *An open question can elicit concrete examples of specific behavior that help an interviewer understand what a client is describing.* |
| | CLIENT: | I would at least like to feel that all the work I do is appreciated and that my wife realizes how much effort I put into helping. |

Choose the most appropriate response.

INTERVIEWER:	But you and your wife have good times, don't you? (Go to 3.42)
INTERVIEWER:	Why doesn't your wife appreciate you and all the work you do? (Go to 3.43)
INTERVIEWER:	What sorts of things would make you feel appreciated? (Go to 3.44)

3.42   Your answer: *But you and your wife have good times, don't you?*
This is a closed question for which you have suggested a specific answer. This response doesn't permit the client to describe his behavior. With an open question, you are more likely to obtain specific information about the client's needs. Return to 3.41 and try again.

3.43	**Your answer:**	*Why doesn't your wife appreciate you and all the work you do?*

This is an open question, but it isn't the most appropriate response. Open questions that begin with the word *why* often irritate clients and make them feel defensive. Appropriate open questions allow clients to express their views and feelings openly. Return to 3.41 and try again.

3.44	**Your answer:**	*What sorts of things would make you feel appreciated?*

Correct. This is an open question that will enable you to obtain specific examples of the client's needs in his present situation.

This interview continues with an open discussion of the client's situation. At its conclusion, the interviewer and the client agree to meet again before any decision is made regarding the appropriateness of counseling sessions. Go to 3.45 to begin the second interview.

3.45

Open questions are very useful as a method of initiating an interview.

CLIENT: (looks inquiringly at the interviewer and waits for the interview to begin)

Choose the most appropriate response.

INTERVIEWER: Were things a little better for you last week? (Go to 3.46)

INTERVIEWER: What would you like to talk about today? (Go to 3.47)

INTERVIEWER: How was your wife last week? (Go to 3.48)

3.46	**Your answer:**	*Were things a little better for you last week?*

This closed question doesn't encourage the client to continue discussing his problem. Return to 3.45 and try again.

3.47	**Your answer:**	*What would you like to talk about today?*

Correct. This open question encourages the client to continue discussing his problem. By beginning with this question, you permit him to take up the discussion where it is most relevant to him. Go to 3.49.

3.48	**Your answer:**	*How was your wife last week?*

Although this is an open question, it focuses on the client's wife rather than on the client. It's important to focus the interview on the client. Return to 3.45 and try again.

3.49

The emphasis in this segment has been on open questions; however, closed questions can sometimes be useful in gaining precise information. They are also helpful in focusing a client on a specific point. The effective interviewer uses both closed and open questions, but uses open questions whenever possible.

CLIENT: I just can't stand living at home anymore. One night last week, I got in the car and felt like smashing it into something and ending it all.

Choose the most appropriate response.

INTERVIEWER: What do you plan to do about your situation? (Go to 3.50)

INTERVIEWER: Why don't you leave home? (Go to 3.51)

INTERVIEWER: Did you actually try to kill yourself? (Go to 3.52)

3.50   **Your answer:**   *What do you plan to do about your situation?*
The client mentioned two important facts. Because of the gravity of his suggestion of self-injury, it is important that you determine whether he is considering suicide. A closed question at this point can help you achieve this objective quickly. If the client is sincere in his desire to end his life, it is essential that you focus on his potential suicide. If the client is not seriously considering suicide, then more will be gained by focusing on his desire to leave home. In either case, you can then revert to an open interviewing style. Return to 3.49 and try again.

3.51   **Your answer:**   *Why don't you leave home?*
This is an open question, but because it sounds like advice, it may cause the client to feel defensive. It's important that you determine whether the client's mention of self-injury or his desire to leave home should be the focus of further discussion. Your response indicates that you have made this decision before obtaining further information from the client. A closed question at this point would assist the client in giving you the required information. Return to 3.49 and try again.

3.52   **Your answer:**   *Did you actually try to kill yourself?*
Correct. This closed question helps you to evaluate the degree of risk associated with the client's present state. If his answer is "yes," his potential suicide should be discussed. If his answer is "no," it would be appropriate to discuss his desire to leave home. In either case, you can then continue to use open questions. Proceed to 3.53.

3.53   *Closed questions can be useful at certain points in an interview. However, don't ask one closed question after another. Be prepared to follow a closed question with a response that allows greater participation on the part of the client.*

CLIENT: No, not really. I could never kill myself.

Choose the most appropriate response.

INTERVIEWER: Could you tell me more about your desire to leave home? (Go to 3.54)

INTERVIEWER: You didn't hurt yourself, then? (Go to 3.55)

INTERVIEWER: You weren't really serious, then, were you? (Go to 3.56)

3.54   **Your answer:**   *Could you tell me more about your desire to leave home?*
Correct. This open question permits the client to discuss his current situation freely. Go to 3.57.

3.55 **Your answer:** *You didn't hurt yourself, then?*
This closed question suggests that you are dissatisfied with the client's response or that you don't believe him. This second closed question may give the client the feeling that he is under cross-examination. Return to 3.53 and try again.

3.56 **Your answer:** *You weren't really serious, then, were you?*
This assumption on your part doesn't encourage the client to discuss his present stressful situation. Your previous response was a closed question; a more open response is required if the client is to become more involved in the discussion. Return to 3.53 and try again.

3.57

Effective use of open and closed questions can result in interviews that are meaningful for both participants. It is possible, however, to misuse this skill during an interview. It's important that you ask only one question at a time. Clients quickly become confused when interviewers ask several questions at once.

CLIENT: As far as I'm concerned, there's nothing I can do to make our marriage work, so I guess I'd be better off out of it. But then, I should stay.

Choose the most appropriate response.

INTERVIEWER: What makes you feel that you should stay? How would you get along if you left? (Go to 3.58)

INTERVIEWER: What important points are entering into your decision? (Go to 3.59)

INTERVIEWER: How would you get along if you left? Would you rent an apartment or move in with a friend? (Go to 3.60)

3.58 **Your answer:** *What makes you feel that you should stay? How would you get along if you left?*
Your initial open question is appropriate, but its impact is lost when you follow it with a second question. Double and multiple questions confuse clients. Return to 3.57 and try again.

3.59 **Your answer:** *What important points are entering into your decision?*
Correct. A single open question facilitates further exploration of the problem by the client. Go to 3.61.

3.60 **Your answer:** *How would you get along if you left? Would you rent an apartment or move in with a friend?*
Often, an open question loses its potential because of the addition of a closed question. Refrain from asking multiple questions. Return to 3.57 and try again.

3.61

When your questions focus on the client's discussion, you avoid leading questions that impose your ideas and assumptions on the client.

CLIENT: Well, if we could get our money problems sorted out, I'm sure my wife and I could get along fine. Before we had money problems, we had good times.

Choose the most appropriate response.

INTERVIEWER: What about getting a consolidated loan? (Go to 3.62)

INTERVIEWER: Could you tell me more about your money problems? (Go to 3.63)

INTERVIEWER: Did you have to acquire so many debts? (Go to 3.64)

3.62   Your answer: *What about getting a consolidated loan?*
Leading questions such as this are inappropriate at any time during an interview. This question assumes that there is a single solution to the client's problem. Return to 3.61 and try again.

3.63   Your answer: *Could you tell me more about your money problems?*
Correct. This open question facilitates discussion of a central difficulty in the client's present situation. Go to 3.65.

3.64   Your answer: *Did you have to acquire so many debts?*
Repeated use of closed questions results in defensiveness on the part of the client, creating a climate that is not conducive to an open discussion. Return to 3.61 and try again.

3.65   *Minimal encouragements indicate to others that you're interested in what they're saying. Once you've asked a question (or used any other interviewing skill), encourage the client to continue talking by using minimal encouragements. A minimal encouragement is a prompt that indicates that you're listening to what the client is saying and that you want him or her to continue.*

CLIENT: We just can't make ends meet. I didn't tell you last week—I couldn't—that I was married before, and I have to make maintenance payments. After I make those payments, we just don't have enough money to live on. That brings me to another problem. I never have time to be with my children . . . and . . . and . . .

Choose the most appropriate response.

INTERVIEWER: How many children do you have? (Go to 3.66)

INTERVIEWER: Do you enjoy your children? (Go to 3.67)

INTERVIEWER: Uhm-hmm. (Go to 3.68)

3.66   Your answer: *How many children do you have?*
The client is telling you about his interactions with his children. He has stopped briefly to gather his thoughts. At this point, a minimal encouragement will indicate that you are interested in what he is saying and would like him to continue. Return to 3.65 and try again.

3.67   Your answer: *Do you enjoy your children?*
This question changes the topic and therefore is not an appropriate response. The client is telling you about his interactions with his children. He has

stopped briefly to gather his thoughts. At this point, a minimal encouragement will indicate that you are interested in what he is saying and would like him to continue. Return to 3.65 and try again.

3.68 **Your answer:** *Uhm-hmm.*

Correct. This minimal encouragement indicates that you are interested in what the client is saying and would like him to continue. Go to 3.69.

3.69

The repetition of one or two key words of a conversation constitutes one type of minimal encouragement. Other examples of minimal encouragements are oh, so, then, uhm-hmm, uh huh, sure, right, *and* go on.

CLIENT: I really would like to . . . well, to have the time to do things with my children.

Choose the most appropriate response.

INTERVIEWER: Would they want that? (Go to 3.70)

INTERVIEWER: Go on. (Go to 3.71)

INTERVIEWER: Do you have boys or girls? (Go to 3.72)

3.70 **Your answer:** *Would they want that?*

This closed question takes the client's attention from what he was telling you and directs it to a topic that interests you. At this point, a minimal encouragement will invite the client to continue with his story. Return to 3.69 and try again.

3.71 **Your answer:** *Go on.*

Correct. This minimal encouragement indicates to the client that you are interested and concerned about what he is saying and would like him to continue. Go to 3.73.

3.72 **Your answer:** *Do you have boys or girls?*

This response changes the topic and interrupts the client. At this point, a minimal encouragement will invite the client to continue with his story. Return to 3.69 and try again.

3.73

Review Questions

The following is a list of closed questions. On a separate sheet, change each into an open question. Try to use a variety of open questions. Compare your responses with the examples that follow the list.

1. Aren't your parents helping you?
2. Do you always argue?
3. Do you try to help?
4. Surely you love your husband?

5. Is it true that you want to leave home?
6. Do you enjoy being with that sort of person?
7. You've never had a partner?
8. Have you stopped fighting with your wife?
9. Shouldn't you consider what your family thinks?
10. Do you want to do something about your problems?

The following open questions would encourage a client to discuss a situation or problem in greater detail.

1. a. Could you tell me how your parents are reacting to your problem?
   b. What kind of support are your parents giving you?
2. a. Could you explain more clearly how you interact?
   b. How do you respond to these situations?
3. a. How do you help?
   b. What do you see as your role?
4. a. What are your feelings about your husband?
   b. Could you tell me about your relationship with your husband?
5. a. How do you feel about leaving home?
   b. Could you clarify for me how you stand on the issue of leaving home?
6. a. What are your feelings about being with that sort of person?
   b. How do you react to being with that sort of person?
7. a. Could you help me to understand your feelings about a permanent relationship?
   b. How do you feel about becoming involved with a partner?
8. a. Could you describe the current situation between you and your wife?
   b. How are you and your wife getting along at present?
9. a. To what degree are you willing to consider your family in this situation?
   b. How important to you are your family's reactions?
10. a. How interested are you in changing your behavior?
    b. To what extent are you willing to work toward changing your behavior?

If you had difficulty providing open questions for three or more of the closed questions, you should work through this chapter again.

Points to Remember about Effective Questioning

1. CONCERNING OPEN QUESTIONS:
   a. Ask a question that cannot be answered with "yes," "no," or a simple fact.
   b. Ask a question that is on topic.
   c. *What* questions are frequently fact oriented.
   d. *How* questions are frequently people oriented.
   e. *Could* and *can* questions provide greatest flexibility for response.

    f. *Why* questions often provoke defensive feelings and are not recommended.
2. ASK OPEN QUESTIONS TO:
    a. Give clients greater opportunity to relate their story.
    b. Gather information and help clients explore and clarify their concerns.
    c. Put clients at ease.
    d. Begin an interview.
    e. Facilitate elaboration of a point.
    f. Elicit specific examples of general situations.
3. WHEN ASKING CLOSED QUESTIONS:
    a. Ask questions that can be answered with "yes," "no," or a simple fact.
    b. Ask questions that are on topic.
4. ASK CLOSED QUESTIONS:
    a. Generally, as infrequently as possible.
    b. Specifically, when you need information that is missing or important to the progress of the interview.
5. FORMS OF QUESTIONS TO BE AVOIDED:
    a. Multiple questions in a single response.
    b. Leading questions.
    c. Cross-examinations.
6. WHEN MAKING MINIMAL ENCOURAGEMENTS:
    a. Use prompts such as *and then, uhm-hmm,* and *right.*
    b. Repeat a few key words from a client's previous statement.
7. MAKE MINIMAL ENCOURAGEMENTS TO:
    a. Encourage a client to continue.
    b. Indicate that you are attending.

3.1

Activity Unit

Find two others who have also read this chapter. Role play brief interviews alternating the client, interviewer, and observer roles. Record each brief interview on either audio- or videotape. An interview consisting of no more than five interviewer responses is sufficient for this exercise.

Interviewer role Practice open questions and minimal encouragements that focus upon and follow the client's communication. Where important to obtain a specific point of information, use a closed question.

Client role Be cooperative and provide sufficient information for the interviewer to practice. Role play the problem of a friend or family member or enact one of the following roles:

1. An individual trying to decide whether or not to begin an exercise program.

Table 3.1

Practice Interview Checklist

	Interviewer response[a]				
Activity	1	2	3	4	5
On topic					
Open question					
Closed question					
Minimal encouragement					
Effective response[b]					

[a] Check the activities that characterize each response.
[b] Indicate the degree of effectiveness on a scale from 1 to 5 in which 1 is ineffective and 5 is very effective.

2. An individual trying to decide whether to buy a house or to continue renting.
3. A person concerned about the future.
4. An individual upset because of the theft of his or her automobile.

Observer role Operate the equipment and complete the Practice Interview Checklist (see Table 3.1).

Postinterview discussion Using the taped interview and the Practice Interview Checklist data, the participants should discuss the performance of the interviewer. All participants can learn from this discussion, which should be nonjudgmental, focus on positive as well as less effective responses, and stimulate improved performance for all.

3.2

Activity Unit

Using the tape of the interview produced in Activity Unit 3.1 or the tape of another practice interview, enter the original interview responses below. Then formulate a better response to replace each original one in the space provided.

Original response 1 _____

_____

_____

Better response _____

_____

_____

Original response 2 _____

_____

_____

Better response _____

_____

_____

Original response 3 _____

_____

_____

Better response _____

_____

_____

Original response 4 _____

_____

_____

Better response _____

_____

_____

Original response 5 _____

_____

_____

Better response _____

_____

_____

Cultural Considerations

1. Remember that question asking is more frequent in the European American culture.
2. Clients from some other cultural groups find direct questions intrusive and lacking in respect. When interviewing clients from these cultures, an indirect

approach such as history taking or engaging clients in a discussion of their lives is a more appropriate way to gather information.

InfoTrac® College Edition Keyword Search Terms

Closed Questions
Effective Questioning
Open Questions
Questions AND Counseling
Types of Questions

Chapter Four

Reflecting Content

This chapter is intended to help you learn to reflect content. When reflecting content, you either paraphrase a single statement or summarize a number of statements. After completing this chapter, you should be able to:

1. Paraphrase a client's statement.
2. Explain the rationale for paraphrasing.
3. Decide whether paraphrasing is appropriate in the situation.
4. Summarize a series of client statements.
5. Explain the rationale for summarizing.
6. Decide whether summarizing is appropriate in the situation.

Reflecting content involves listening carefully to a client and feeding back the essence of what the client has said more concisely in the interviewer's own words. A *paraphrase* reflects the content of what a client has just communicated. A *summarization* reflects the content of a number of client statements, covering more information and a longer time span. The purpose of a paraphrase is to encourage a client to discuss an issue in more detail and to ensure that the communication has been understood by both the client and interviewer. Summarizations are used to tie together multiple elements of client messages, to identify a common theme that has emerged over a number of statements or even interviews, to focus a rambling client, to moderate the pace of an interview, and to review progress over one or more interviews.

Interviewer trainees sometimes confuse the restatement of the content or cognitive aspects of a client's statement with the repetition of a client's message, or "parroting." When parroting occurs, interviews become circular rather than progressive, and clients become uncomfortable and find it difficult to continue. Therefore, pay careful attention to the essence of a client's statement rather than to the words a client uses.

The program in this chapter focuses on an interview that takes place in the personnel office of an insurance firm. The client—an insurance agent who is married and has one daughter—has been employed by the firm for about 20 years. Until recently, when his work began to deteriorate, he was regarded as a reliable and efficient employee. An interview has been arranged by the personnel manager to discuss the employee's tardiness and absenteeism. Go to 4.1.

4.1		*To reflect the content of a client's statement, an interviewer must paraphrase the main idea contained in the statement.*
	CLIENT:	I'm not really sure why you've asked me to see you today, unless it's because I've been late a few times this month.
		Choose the response that best reflects the content of the client's comments.
	INTERVIEWER:	I'm sure you know why I've called you in to see me. (Go to 4.2)
	INTERVIEWER:	You think that your tardiness is causing problems. (Go to 4.3)
	INTERVIEWER:	We do have to talk about your tardiness (Go to 4.4)

4.2   **Your answer:**   *I'm sure you know why I've called you in to see me.*
This response fails to acknowledge the content of the client's statement. It detracts from his comment and indicates that you weren't focusing on what he was saying. Try to identify the essence of the client's statement. Return to 4.1 and try again.

4.3   **Your answer:**   *You think that your tardiness is causing problems.*
Correct. You have paraphrased the essence of the client's communication without changing its meaning. Go to 4.5.

4.4	**Your answer:**	*We do have to talk about your tardiness.*

We do have to talk about your tardiness.
Your response acknowledges the fact that you've heard the client, but it significantly detracts from his communication and will probably increase his apprehension. Return to 4.1 and choose a response that focuses on the main thought expressed by the client.

4.5

As an interviewer, you should reflect the essence of a client's communication, no matter where it occurs in his or her response.

CLIENT: Well, it's one of the problems I'm concerned about, and I know you people worry about attendance and punctuality.

Choose the response that best reflects the content of the client's comments.

INTERVIEWER: There are a number of problems bothering you right now. (Go to 4.6)

INTERVIEWER: You seem to be aware of my position as personnel manager. (Go to 4.7)

INTERVIEWER: Then you know that the firm is interested in your attendance and punctuality. (Go to 4.8)

4.6   Your answer: *There are a number of problems bothering you right now.*
Correct. This reflection of content focuses on the most important aspect of the client's comment. Proceed to 4.9.

4.7   Your answer: *You seem to be aware of my position as personnel manager.*
This is an inappropriate reflection of content. The essence of a client's response isn't always contained in his or her last comment. Be concerned with the client's problems, not his knowledge of your position. Return to 4.5 and try again.

4.8   Your answer: *Then you know that the firm is interested in your attendance and punctuality.*
You've parroted the last part of the client's comment and ignored the message contained in his response. Although your response isn't totally inappropriate, it fails to help you or the client explore his problem in any depth; instead, it concentrates on the interests of the firm. Return to 4.5 and try again.

4.9

By appropriately reflecting content, you assure clients that you understand their story.

CLIENT: I guess all my problems are caused by what's going on at home.

Choose the response that best reflects the content of the client's comments.

INTERVIEWER: You have a problem with your marriage. (Go to 4.10)

INTERVIEWER: You think that your problems stem from your difficulties at home. (Go to 4.11)

INTERVIEWER: You don't seem to be concerned about your status in the firm. (Go to 4.12)

4.10   **Your answer:**   *You have a problem with your marriage.*
You've assumed too much. With this reflection of content, you've gone beyond the essence of what the client has said. He may have a number of problems at home. It's unwise to assume that you know what his problem is before he tells you about it. If he has no problem with his marriage, the client is likely to react negatively to your comment. Return to 4.9 and try again.

4.11   **Your answer:**   *You think that your problems stem from your difficulties at home.*
Correct. By accurately reflecting the content of his last statement, you've confirmed that you've heard the client's communication and that you want him to continue. Go to 4.13.

4.12   **Your answer:**   *You don't seem to be concerned about your status in the firm.*
Your response gives no indication to the client that you've heard his communication. You've ignored the essence of his message. Return to 4.9 and try again.

4.13

A reflection of content may take the form of a simple, compound, or complex sentence, or a sentence fragment. The reflection should be similar in grammatical structure to the client's statement.

CLIENT: Yes. It's my daughter. She's 15, you know, and we're having trouble with her. She stays out all night.

Choose the response that best reflects the content of the client's comments.

INTERVIEWER: Has your daughter ever been in trouble with the law? (Go to 4.14)

INTERVIEWER: I guess you've had a lot of problems with your daughter. (Go to 4.15)

INTERVIEWER: The fact that your daughter doesn't come home at night is a major problem. (Go to 4.16)

4.14   **Your answer:**   *Has your daughter ever been in trouble with the law?*
This is a closed question focusing on the client's daughter rather than on the client and his problems. Your reflection should be grammatically similar to the client's comment; therefore, you should respond with a statement, not a question. Return to 4.13 and try again.

4.15   **Your answer:**   *I guess you've had a lot of problems with your daughter.*
You've assumed too much and changed the focus from a discussion of the client's present difficulty with his daughter to his past experience with her. Return to 4.13 and try again.

4.16   **Your answer:**   *The fact that your daughter doesn't come home at night is a major problem.*
Correct. This reflection of content paraphrases the client's comments and, like his response, is in the present tense. Go to 4.17.

4.17

In reflecting content, emphasize the cognitive aspect of a client's remarks.

CLIENT: That, and the effect it's having on my wife's heart condition. She's already had one heart attack.

Choose the response that best reflects the content of the client's comments.

INTERVIEWER: So, you also have your wife's health to consider. (Go to 4.18)

INTERVIEWER: You're really worried about your wife. (Go to 4.19)

INTERVIEWER: Your daughter must be a real problem for your wife. (Go to 4.20)

4.18 Your answer: *So, you also have your wife's health to consider.*
Correct. This reflection of content focuses on the cognitive aspect of the client's comments—his consideration of his wife's health. Proceed to 4.21.

4.19 Your answer: *You're really worried about your wife.*
This response is a good reflection of feeling; however, you were asked for a reflection of content. Return to 4.17 and try again.

4.20 Your answer: *Your daughter must be a real problem for your wife.*
Although this response relates to the content of the client's comments, it shifts the focus of the interview from his problems to problems his wife may be having. Return to 4.17 and try again.

4.21

A reflection of content can summarize an interview and pull together the essence of a number of the client's statements.

CLIENT: I know I'm having problems at work, but what am I going to do?

Choose the response that best summarizes the client's comments to this point.

INTERVIEWER: I understand that you have problems at home, but can we focus on your attendance problems here? (Go to 4.22)

INTERVIEWER: Now let me see if I understand your problem. You realize that you're having difficulties at work, but as you understand it, these are related to your problems at home. (Go to 4.23)

INTERVIEWER: It seems that you're finding it difficult to do your job satisfactorily and, at the same time, cope with the problems caused by your daughter and your wife's health. (Go to 4.24)

4.22 Your answer: *I understand that you have problems at home, but can we focus on your attendance problems here?*
This response is too brief and abrupt. The client has revealed three problems, and you should encourage further open discussion of these problems. A good summarization gives equal emphasis to the various themes covered in the client's previous statements, allowing the client to select the problems to be discussed further. Return to 4.21 and try again.

4.23   **Your answer:**   *Now let me see if I understand your problem. You realize that you're having difficulties at work, but as you understand it, these are related to your problems at home.* You've focused on the apparent problems of the client rather than on his actual problems. Return to 4.21 and try again.

4.24   **Your answer:**   *It seems that you're finding it difficult to do your job satisfactorily and, at the same time, cope with the problems caused by your daughter and your wife's health.* Correct. This summary accurately reflects the essence of the client's story up to this point in the interview. Go to 4.25.

4.25

Accurate reflection of content encourages and reassures clients who are threatened by discussing their feelings.

CLIENT: Yes, but the major difficulty is with my daughter. I think she's been sleeping around, and I'm at my wits' end to know what to do about it.

Choose the most appropriate response (assuming that this client has difficulty talking about his feelings).

INTERVIEWER: It sounds as though you feel desperate about the situation. (Go to 4.26)

INTERVIEWER: That must be an awful situation. (Go to 4.27)

INTERVIEWER: Your major concern, then, is what to do to help your daughter. (Go to 4.28)

4.26   **Your answer:**   *It sounds as though you feel desperate about the situation.* This is an accurate reflection of feeling; however, the client is not yet in touch with his feelings, and your reflection at this point could be threatening to him. If the client does feel threatened, he may decide that he's telling you too much too soon and that he should avoid this particular topic. Return to 4.25 and try again.

4.27   **Your answer:**   *That must be an awful situation.* This is an accurate reflection of feeling; however, you were asked to assume that the client has difficulty discussing feelings. Whenever the client finds it difficult to discuss feelings, you should reflect content and gradually introduce reflections of feeling when the client becomes comfortable talking about feelings. Return to 4.25 and try again.

4.28   **Your answer:**   *Your major concern, then, is what to do to help your daughter.* Correct. This reflection of content is the most appropriate response for this client. Go to 4.29.

4.29

By reflecting content, you communicate to the client that you are following and attempting to understand what is being said.

CLIENT: Yes. You see, we've tried everything. We've promised to give her things if she'll agree to stay home. We've tried to ground her. What else can I do?

Choose the response that best reflects the content of the client's comments.

INTERVIEWER: You've tried a lot of things, but nothing has worked. (Go to 4.30)

INTERVIEWER: Well, maybe your minister could help you. (Go to 4.31)

INTERVIEWER: You think you should do more. There must be something else you can do. (Go to 4.32)

4.30 **Your answer:** *You've tried a lot of things, but nothing has worked.*
Correct. This reflection of content indicates that you're attempting to under-stand the message the client is conveying to you. If there is a misunderstand-ing, this will give him an opportunity to clarify what he has said. Go to 4.33.

4.31 **Your answer:** *Well, maybe your minister could help you.*
This advice changes the topic and indicates to the client that you aren't focus-ing on what he is saying. Return to 4.29 and try again.

4.32 **Your answer:** *You think you should do more. There must be something else you can do.*
This reflection of content ignores the essence of the client's message. He has just told you about his attempts to help his daughter and his failure to do so; your reflection of content should recognize this fact, not belittle it. Return to 4.29 and try again.

4.33

The main purpose of summarizing content is to help clients collect their thoughts. Clients often present their ideas in a haphazard way. To help clients organize their concerns, interviewers should present a concise, accurate, and timely summary.

CLIENT: I lost my temper and almost hit her. But my wife got so upset, I started worrying about her heart condition. Anyway, I realized that I wouldn't solve anything by losing my temper.

Choose the response that best summarizes the client's comments up to this point.

INTERVIEWER: So, your problems at work stem from the problems you're having with your daughter. (Go to 4.34)

INTERVIEWER: So, the stress caused by your wife's heart condition and your daughter's promiscu-ity is affecting your ability to work, and you're searching for a solution. (Go to 4.35)

INTERVIEWER: You're aware that your work isn't what it used to be. You seem to think your problems with your daughter are contributing to your problems at work and put-ting a considerable strain on your wife. Your daughter is becoming promiscuous and, despite your efforts to control her, she continues to act out. (Go to 4.36)

4.34 **Your answer:** *So, your problems at work stem from the problems you're having with your daughter.*
As an interviewer, you should focus on the client's main problem. This sum-mary reflects only a small portion of the client's discussion and fails to help him organize his disclosures. Return to 4.33 and try again.

4.35   Your answer: *So, the stress caused by your wife's heart condition and your daughter's promiscuity is affecting your ability to work, and you're searching for a solution.*
Correct. Your response is brief, accurate, and well organized. It will help the client deal with his problems in an orderly way. Go to 4.37.

4.36   Your answer: *You're aware that your work isn't what it used to be. You seem to think your problems with your daughter are contributing to your problems at work and putting a considerable strain on your wife. Your daughter is becoming promiscuous and, despite your efforts to control her, she continues to act out.*
Although this reflection of content contains the essence of what the client has discussed, it is fragmented. A summary should be brief, well organized, and to the point. Return to 4.33 and try again.

4.37   *Reflection of content can provide direction for an interview.*

CLIENT: I guess I'm going to have to find some help in dealing with my daughter's problem. I suppose I should be discussing my work-related problems.

Choose the response that best reflects the content of the client's comments.

INTERVIEWER: I wonder if your daughter really needs help. (Go to 4.38)

INTERVIEWER: Yes. Your problems here at work should be the focus of our discussion. (Go to 4.39)

INTERVIEWER: Your daughter's problem is important, but your problems at work are important, too. (Go to 4.40)

4.38   Your answer: *I wonder if your daughter really needs help.*
This reflection of content is inaccurate. It shifts the focus of the interview from the client's concerns to your thoughts. Moreover, this response focuses on only part of the client's statement. When you reflect content, focus on main themes so that the direction of the interview is clear. Return to 4.37 and try again.

4.39   Your answer: *Yes. Your problems here at work should be the focus of our discussion.*
You've ignored an important portion of the client's message—his problems at home. Return to 4.37 and try again.

4.40   Your answer: *Your daughter's problem is important, but your problems at work are important, too.*
Correct. By concisely reflecting the content of the client's statement, you've made the direction of the interview clear to him. Proceed to 4.41.

4.41   *By reflecting content, you can check and clarify your perceptions of what a client has said. Before you can help a client, you need to understand what he or she is trying to say to you.*

CLIENT: Well, since I found out about my daughter's problem, I've spent more time at home, and we've had more and more late-night arguments. As a result, I can't get up early in the morning, so I'm late. Then sometimes I just don't feel like coming to work.

Choose the response that best reflects the content of the client's comments.

INTERVIEWER: It sounds as though you've lost interest in your work. (Go to 4.42)

INTERVIEWER: Your problems with your daughter have been interfering with your work. (Go to 4.43)

INTERVIEWER: It sounds as though your wife should take more responsibility for what's going on at home. (Go to 4.44)

4.42 **Your answer:** *It sounds as though you've lost interest in your work.*

This response doesn't paraphrase the client's comments. You've assumed, perhaps incorrectly, that he has lost interest in his work. Return to 4.41 and try again.

4.43 **Your answer:** *Your problems with your daughter have been interfering with your work.*

Correct. By reflecting the content of the client's comments, you've clarified your own perceptions of his remarks. If your perceptions are incorrect, the client will supply you with additional information. Go to 4.45.

4.44 **Your answer:** *It sounds as though your wife should take more responsibility for what's going on at home.*

This may or may not be true. In either case, this response will be of little benefit to the client, who is trying to tell you about his problem. Your responsibility at this point is to check and clarify your perceptions of what the client has told you. Return to 4.41 and try again.

4.45

Reflection of content should be interspersed with other types of interviewer responses. The overuse of reflecting has a parrotlike effect that can inhibit a client's communication.

CLIENT: Yes. And even when I try to keep my mind on my work, I find that my attention drifts back to the problems I'm having with my daughter. I've lost a few accounts lately. I lost some of them because I just couldn't get involved enough. And the others . . . well, there were times when my wife called, and I just had to break appointments and go home.

Choose the response that best describes appropriate interviewer behavior.

INTERVIEWER: (having previously responded with an open question) Even when you're at work, your problems at home interfere with what you're doing. (Go to 4.46)

INTERVIEWER: (having previously responded with seven reflections of content) Even when you're at work, your problems at home interfere with what you're doing. (Go to 4.47)

INTERVIEWER: (having reflected content for 5 minutes) Even when you're at work, your problems at home interfere with what you're doing. (Go to 4.48)

4.46   Your answer: *(having previously responded with an open question) Even when you're at work, your problems at home interfere with what you're doing.*
Correct. This response indicates your awareness of the fact that reflection of content shouldn't be overused. Go to 4.49.

4.47   Your answer: *(having previously responded with seven reflections of content) Even when you're at work, your problems at home interfere with what you're doing.*
Continuous reflection of content will give the client the impression that you are simply parroting what he is saying. Return to 4.45 and try again.

4.48   Your answer: *(having reflected content for 5 minutes) Even when you're at work, your problems at home interfere with what you're doing.*
Reflection of content should be interspersed with other types of responses if communication between yourself and the client is to be facilitated. Return to 4.45 and try again.

4.49   *Summarization of content is a systematic integration of the important themes in a client's story, not a mechanical means of pulling together a number of facts. It is important to listen carefully to recall accurately the major themes discussed by the client. Summarization helps a client review what has been said and stimulates a thorough exploration of themes discussed.*

CLIENT: Just the other day, I heard that one of my best clients is moving his insurance business to the Comprehensive Life Insurance Company.

Choose the response that best summarizes the client's story up to this point.

INTERVIEWER: It seems that your tardiness, your frequent absence, and your inability to attend to your work have lost the company considerable business. (Go to 4.50)

INTERVIEWER: It seems that, at the moment, your problems at home take precedence over your job. (Go to 4.51)

INTERVIEWER: You seem to be aware of the fact that the problems you're having at home are having serious effects on your business activities. (Go to 4.52)

4.50   Your answer: *It seems that your tardiness, your frequent absence, and your inability to attend to your work have lost the company considerable business.*
Throughout the interview, the client has related his problems at home to his problems at work. This response will indicate to him that you've heard only his comments about his job. This inaccurate summarization of the client's comments represents your own perceptions rather than those of the client. Return to 4.49 and try again.

4.51   Your answer: *It seems that, at the moment, your problems at home take precedence over your job.*
You've attempted to summarize the main facts in the client's story, but your integration of these facts has resulted in a misinterpretation of the underlying

themes. The client has indicated concern for his business competence as well as an awareness of the relationship between his problems at home and his problems at work. Return to 4.49 and try again.

4.52   **Your answer:**   *You seem to be aware of the fact that the problems you're having at home are having serious effects on your business activities.*
Correct. This brief summary reviews the essential ideas that are contained in the client's story, promotes further exploration, and will lead toward solution of his problems. Go to 4.53.

4.53

You can bring an interview to an effective conclusion by summarizing the major themes covered during the session.

CLIENT: Speaking of business, I have an appointment in a few minutes.

Choose the response that best concludes the interview.

INTERVIEWER: I guess you'd better go. You seem really concerned about this business appointment. (Go to 4.54)

INTERVIEWER: I guess you'd better go. What do you think you should do about your problem? (Go to 4.55)

INTERVIEWER: I guess you'd better go. Let's see . . . we've discussed your family problems, their effect on your business activities, and your need to find solutions. (Go to 4.56)

4.54   **Your answer:**   *I guess you'd better go. You seem really concerned about this business appointment.*
This reflection of feeling will probably initiate discussion of the client's appointment rather than close the interview. A brief summarization is often a good way to close down an interview. It allows a mutual review of the themes covered and sets the scene for the next interview. Return to 4.53 and try again.

4.55   **Your answer:**   *I guess you'd better go. What do you think you should do about your problem?*
The client has indicated that he has a business appointment. As personnel manager, you should realize the importance of his appointments, and your response should serve to review and terminate the interview. Return to 4.53 and try again.

4.56   **Your answer:**   *I guess you'd better go. Let's see . . . we've discussed your family problems, their effect on your business activities, and your need to find solutions.*
Correct. This brief reflection of content summarizes the essence of the interview and sets the scene for the next interview. Go to 4.57.

4.57

After you've investigated the client's need to discuss any problems that have come up since your last meeting, briefly summarize that meeting to give direction to the present interview.

INTERVIEWER: If you've experienced any major difficulties since our last meeting, maybe we should discuss these first.

CLIENT:	Well, not really. Things are much the same as they were yesterday.

Choose the response that best initiates and gives direction to the interview.

INTERVIEWER: Yesterday, you discussed some of your problems at home and their effects on your work. You also discussed your need to solve these problems. (Go to 4.58)

INTERVIEWER: The major concern we dealt with yesterday is the fact that your personal problems are beginning to affect your work. (Go to 4.59)

INTERVIEWER: Yesterday, we were discussing the problems you're having at home. You indicated that you think your daughter's problem is serious, and the question of whether she needs professional care came up. (Go to 4.60)

4.58   Your answer: *Yesterday, you discussed some of your problems at home and their effects on your work. You also discussed your need to solve these problems.*

Correct. This brief, accurate summary reviews your last session and enables the client to resume his discussion. Go to 4.61.

4.59   Your answer: *The major concern we dealt with yesterday is the fact that your personal problems are beginning to affect your work.*

This may be your major concern, but it isn't the client's. By focusing on your major concern, you're distorting what the client has told you. As personnel manager, you can best help the client by accurately summarizing his major disclosures. Only when you understand the nature of his problem will you be able to help him resolve it. Return to 4.57 and try again.

4.60   Your answer: *Yesterday, we were discussing the problems you're having at home. You indicated that you think your daughter's problem is serious, and the question of whether she needs professional care came up.*

You've added your own ideas to those discussed by the client. When you reflect content in the form of a summary, try to be accurate and avoid adding to or changing the client's message by focusing on your own ideas. Return to 4.57 and try again.

4.61   *Summarization can be used to clarify a client's confusing, lengthy, or rambling comments.*

CLIENT: My sister took her daughter to their family physician, and he had a long chat with her.

INTERVIEWER: Your sister found help.

CLIENT: Yes. I still don't know whether I should see the school counselor about my daughter. It might be a good idea. I just don't know.

INTERVIEWER: It's difficult to know what to do.

CLIENT: John, across the road, said something about getting her into a detention home, but that's kind of drastic. I don't know whether there is any solution. Maybe I could send her to live with my brother in Los Angeles.

Choose the most appropriate response.

INTERVIEWER: I'm sure there are other solutions. (Go to 4.62)

INTERVIEWER: Your sister sent her daughter to the family doctor. You wonder about taking your daughter to see a school counselor. Your neighbor has suggested a detention home. And you've thought of involving your brother. (Go to 4.63)

INTERVIEWER: It seems that you're thinking about a number of solutions to your problem, and you're not sure which solution you should choose. Maybe we can discuss each one separately. (Go to 4.64)

4.62 **Your answer:** *I'm sure there are other solutions.*
You've ignored the client's efforts to solve his problems. By suggesting that his thoughts on the matter are limited, you've failed to help him investigate the solutions he has thought of. Return to 4.61 and try again.

4.63 **Your answer:** *Your sister sent her daughter to the family doctor. You wonder about taking your daughter to see a school counselor. Your neighbor has suggested a detention home. And you've thought of involving your brother.*
When you reflect content, be brief and well organized and focus on the essential ideas contained in the client's story. This rambling, parrotlike reflection of content misses the essence of what the client has communicated to you. Return to 4.61 and try again.

4.64 **Your answer:** *It seems that you're thinking about a number of solutions to your problem, and you're not sure which solution you should choose. Maybe we can discuss each one separately.*
Correct. This summary helps the client identify the meaning behind his rambling thoughts and directs him toward the next phase of the interview—exploration of possible alternatives. Summarization can help a client explore and set goals. Go to 4.65.

4.65

Review Questions

Indicate whether each of the following statements is true or false.

1. When reflecting content, it's important to repeat accurately everything a client has said.
2. Effective reflections of content confirm for the client that his or her message has been understood.
3. Frequent use of reflection of content may cause a parrotlike effect and inhibit a client's communication.
4. When clients are threatened by discussion of their feelings, the interviewer can use reflections of content.
5. In framing a reflection of content, an interviewer should concentrate on a client's most recent remarks because these are usually most important.

6. Summarization is appropriate when a client's comments have been confusing, lengthy, or rambling.
7. When summarizing, an interviewer must include all of the topics discussed by a client.
8. Summarization is inappropriate at the conclusion of an interview.
9. Reflection of content should be concise and should capture the essence of a client's comments.
10. Summarization may be used to give direction to an interview.

Review Answers

1. *False.* If you answered *true*, go to 4.1 and review.
2. *True.* If you answered *false*, go to 4.9 and review.
3. *True.* If you answered *false*, go to 4.45 and review.
4. *True.* If you answered *false*, go to 4.25 and review.
5. *False.* If you answered *true*, go to 4.5 and review.
6. *True.* If you answered *false*, go to 4.61 and review.
7. *False.* If you answered *true*, go to 4.49 and review.
8. *False.* If you answered *true*, go to 4.53 and review.
9. *True.* If you answered *false*, go to 4.33 and review.
10. *True.* If you answered *false*, go to 4.57 and review.

If seven or more of your responses are wrong, you should return to 4.1 and review this chapter.

Points to Remember about Paraphrasing

1. WHEN PARAPHRASING:
   a. Restate the main ideas contained in a client's communication.
   b. Don't add to or change the meaning of a client's statement.
   c. Avoid parroting a client's comments.
2. PARAPHRASING:
   a. Indicates that you're attending to and attempting to understand what a client is saying.
   b. Helps develop a working relationship between you and the client.
   c. Serves to check your understanding of the client's statement.
   d. Helps the client crystallize his or her thoughts.
   e. Gives direction to an interview.
3. YOU CAN PARAPHRASE:
   a. When a client is threatened by discussion of feelings.
   b. To check and clarify your perceptions of what a client is saying.
   c. To indicate to a client that you understand what has been said, facilitating further discussion.

Points to Remember about Summarizing

1. WHEN SUMMARIZING:
   a. Systematically integrate the important ideas contained in a client's story and restate them.
2. SUMMARIZING:
   a. Provides concise, accurate, and timely overviews of clients' narratives and helps them reorganize their thoughts.
   b. Helps a client review what has been said.
   c. Stimulates a thorough exploration of themes that are important to the client.
   d. Provides organization for an interview.
3. YOU CAN SUMMARIZE:
   a. When a client's comments are lengthy, rambling, or confused.
   b. When a client presents a number of unrelated ideas.
   c. To add direction and coherence to an interview.
   d. To move from one phase of an interview to the next.
   e. To conclude an interview.
   f. To provide an introduction to an interview by reviewing the previous interview.

4.1

Activity Unit

Find two others who have also read this chapter. Role play brief interviews (no more than five interviewer responses) alternating the client, interviewer, and observer roles. Record each brief interview on audio- or videotape.

Interviewer role Practice paraphrasing the client's statements. The last response in the series should summarize the client's statements.

Client role Be cooperative and provide sufficient information for the interviewer to practice. Clients may want to role play the problem of a friend or family member or to enact one of the following roles:

1. An individual who has so many things to do at work or home that no time is left to engage in and enjoy leisure activities.
2. A parent who adopted a child 2 years ago and is having second thoughts about it.
3. An individual who is concerned about nuclear proliferation.
4. A person who has just inherited a considerable amount of money and property and cannot decide what to do with it.

| Table 4.1 |

Practice Interview Checklist

Activity	Interviewer response[a]				
	1	2	3	4	5
On topic					
Paraphrase					
Summarization					
Effective response[b]					

[a] Check the activities that characterize each response.
[b] Indicate the degree of effectiveness on a scale from 1 to 5 in which 1 is ineffective and 5 is very effective.

Observer role Operate the equipment and complete the Practice Interview Checklist (see Table 4.1).

Postinterview discussion Using the taped interview and the Practice Interview Checklist data, the participants should discuss the performance of the interviewer. All participants can learn from this discussion, which should be nonjudgmental, focus on positive as well as less effective responses, and stimulate improved performance for all.

4.2   Activity Unit

Using the tape of the interview produced in Activity Unit 4.1 or the tape of another practice interview, enter the original interview responses below. Then formulate a better response to replace each original one in the space provided.

Original response 1 _____

_____

_____

Better response _____

_____

_____

Original response 2 _____

_____

_____

Better response _____

_____

_____

Original response 3 _____

_____

_____

Better response _____

_____

_____

Original response 4 _____

_____

_____

Better response _____

_____

_____

Original response 5 _____

_____

_____

Better response _____

_____

_____

Cultural Considerations

1. When reflecting content with clients from other cultures, it is important to consider the client's perception of the world as developed by his or her family or cultural experience.
2. Take particular care that you listen carefully and understand the clients' communications when they speak with an accent different from yours, or use jargon or colloquialisms with which you are unfamiliar.

InfoTrac® College Edition Keyword Search Terms

Paraphrasing AND Communication
Paraphrasing AND Counseling
Paraphrasing AND Interviewing
Summarizing AND Communication
Summarizing AND Interviewing

Chapter Five

Reflecting Feeling

*T*his chapter is intended to help you master the skill of reflecting feeling. After completing this chapter, you should be able to:

1. Identify the feelings a client conveys to you by selecting those words that best describe them.
2. Reflect a client's feelings with words that identify his or her emotions.
3. State the rationale for reflecting feeling.
4. Decide whether reflection of feeling is appropriate in the situation.

Reflecting feeling involves identifying client emotions and then rephrasing this affective component back to the client. The feelings may be found in the client's direct communication or inferred from either the client's non-verbal behavior, or the context or content of the client's communication. When an interviewer reflects feelings, clients are assisted to explore their feeling in greater detail, to understand their sometimes hidden emotions, or to sort out complex and confusing emotions. Reflecting feeling ensures that the affective component of the client's communication is clear to both the client and the interviewer. Effective and accurate reflection of feeling can indicate to clients that they are understood, facilitating the client's open communication with the interviewer.

The first part of this chapter deals with relatively simple emotions in order to help you develop the ability to identify and reflect feelings. In actual practice, however, individuals rarely convey simple emotions when they communicate. Most people display multiple emotions when they interact with each other; therefore, the latter part of this chapter is devoted to the identification and reflection of "mixed" emotions. When identifying the emotions of others, it is important to attend to not only *what* is said but also *how* it is said. The posture, voice, and mannerisms of individuals often provide important information about their emotions.

To help you develop skills in identifying and reflecting feeling, this chapter follows a different format. Excerpts from interviews with a variety of clients permit presentation of a more varied range of emotions than is possible in the single-case format. Each frame is complete in itself; that is, each depicts a different client. Go to 5.1.

5.1		*To reflect feeling, you must identify the emotions a client is experiencing. This first series of frames helps you develop the ability to identify emotions.*

CLIENT: (describing his relationship with his employer) I try and I try, but I hardly ever seem to succeed. Every time I try to do what he wants, it doesn't work out. When I try to do things the way I think they should be done, he doesn't like that either. I just don't know what to do.

Choose the word that best represents the client's feelings.

INTERVIEWER: Guilty (Go to 5.2)

INTERVIEWER: Angry (Go to 5.3)

INTERVIEWER: Frustrated (Go to 5.4)

5.2   Your answer: *Guilty*
The client's response indicates that he is trying and has tried to do something about his situation. He would be more likely to feel guilty if he hadn't tried to do something about it. Return to 5.1 and try again.

5.3   Your answer: *Angry*
The client may be angry, but you need more information to substantiate this conclusion. To identify feelings accurately, you must attend closely to the client. Return to 5.1 and try again.

5.4   Your answer: *Frustrated*
Correct. The client is trying to reach his goal, but his attempts have failed and he feels frustrated. Go to 5.5.

5.5 CLIENT: (whose car was recently damaged in an accident) I'd like to take something and wrap it around his head—anything to get even with that bastard.

Choose the word that best describes the client's feelings.

Depressed (Go to 5.6)

Revengeful (Go to 5.7)

Annoyed (Go to 5.8)

5.6 **Your answer:** *Depressed*

The client is feeling hostile—not depressed. Return to 5.5 and try again.

5.7 **Your answer:** *Revengeful*

Correct. Although the first part of the client's statement communicates anger, the second part communicates a desire to get even. Go to 5.9.

5.8 **Your answer:** *Annoyed*

The client is undoubtedly annoyed; however, the language being used suggests feelings far stronger than mere annoyance. This response is an understatement of the client's feelings. Return to 5.5 and try again.

5.9 CLIENT: (discussing a close friend) It bothers me, and I really worry about him. I want to help, but I just can't get through to him.

Choose the emotion that best represents the client's feelings.

Resigned (Go to 5.10)

Frustrated (Go to 5.11)

Concerned (Go to 5.12)

5.10 **Your answer:** *Resigned*

When attempting to identify feelings, you must attend to the client's entire message. The client's expressed desire to help doesn't indicate an attitude of resignation. Return to 5.9 and try again.

5.11 **Your answer:** *Frustrated*

Although the client may be frustrated, the feelings being revealed can be more accurately identified. Return to 5.9 and try again.

5.12 **Your answer:** *Concerned*

Correct. "I want to help . . ." suggests concern. Go to 5.13.

5.13 CLIENT: (discussing her recent remarriage) There is just one feeling I have when I look at him; I'm not sure I can find the word. I feel good inside—sort of glowing—like I used to when I woke up on Christmas morning.

Choose the emotion that best represents the client's feelings.

Joyous (Go to 5.14)

Surprised (Go to 5.15)

Appreciated (Go to 5.16)

5.14   Your answer: *Joyous*

Correct. It's as important to identify positive emotions as it is to identify negative ones. The client's statement suggests a strong positive emotion. Go to 5.17.

5.15   Your answer: *Surprised*

The client hasn't said her new relationship surprises her; she has told you how good and glowing she feels inside. Return to 5.13 and try again.

5.16   Your answer: *Appreciated*

The client may feel appreciated, but she hasn't mentioned this feeling. She focuses on her own internal feelings rather than on her feelings about her relationship with her husband. Return to 5.13 and try again.

5.17   CLIENT: (whose wife and children have recently left him) I feel deserted. There's nowhere to turn—just nowhere. I feel so left out. I'm sure no one cares whether I live or die.

Choose the emotion that best represents the client's feelings.

Depressed (Go to 5.18)

Apprehensive (Go to 5.19)

Afraid (Go to 5.20)

5.18   Your answer: *Depressed*

Correct. The first part of the client's statement indicates loneliness. The second part indicates that he believes no one cares whether he lives or dies—an indication of depression. Go to 5.21.

5.19   Your answer: *Apprehensive*

The client may be apprehensive; however, his words indicate that the emotion he is feeling is stronger than apprehension. Return to 5.17 and try again.

5.20   Your answer: *Afraid*

It is likely that the client is frightened by his situation; however, his last comment indicates that his reaction constitutes more than fear. Return to 5.17 and try again.

5.21		In frames 5.1 through 5.20, you learned how to identify feelings. One of the most important aspects of helping others is developing the ability to reflect the identified feeling back to the client. The following frames help you learn how to form appropriate reflections of feelings.
	CLIENT:	(describing her husband's reaction to her decision to find a job) He laughed at me. My own husband just sat there and laughed at me. I felt like such a fool—so put down.
		Choose the response that best reflects the client's feelings.
	INTERVIEWER:	Why didn't you tell him to stop? (Go to 5.22)
	INTERVIEWER:	You felt humiliated. (Go to 5.23)
	INTERVIEWER:	You felt angry. (Go to 5.24)

5.22 **Your answer:** *Why didn't you tell him to stop?*
This is an open question rather than a reflection of feeling. Return to 5.21 and try again.

5.23 **Your answer:** *You felt humiliated.*
Correct. This is an accurate reflection of the client's description of her feelings. It's often helpful to imagine yourself in the situation the client describes—you would probably feel humiliated if someone laughed at you. Go to 5.25.

5.24 **Your answer:** *You felt angry.*
You should avoid overinterpretation of the client's feelings. This response assumes too much. The situation the client has described may have produced anger, hurt, or even depression; however, the client's response suggests that she reacted differently. Return to 5.21 and try again.

5.25		When reflecting feeling, notice the affective words used in an interview and reflect them, or some variation of them, back to the client without repeating the client's words.
	CLIENT:	(describing difficulties at work) I try, honestly, but I don't get anywhere. Working hard doesn't seem to make any difference; I'm still behind.
		Choose the response that best reflects the client's feelings.
	INTERVIEWER:	I guess you're depressed. (Go to 5.26)
	INTERVIEWER:	I guess you must feel that working hard doesn't get you anywhere. (Go to 5.27)
	INTERVIEWER:	I guess you feel discouraged. (Go to 5.28)

5.26 **Your answer:** *I guess you're depressed.*
This response is a reflection of feeling, but it is too heavily loaded. The client has not indicated that she is depressed, only that she is discouraged.

A response that is too strong may threaten or irritate the client. Return to 5.25 and try again.

5.27   Your answer: *I guess you must feel that working hard doesn't get you anywhere.*
In making this response, you rely too heavily on the client's words. When you take a client's words out of context and repeat them, you sometimes change their meaning. Return to 5.25 and try again.

5.28   Your answer: *I guess you feel discouraged.*
Correct. This is an accurate reflection of feeling in which you identify the affect associated with the client's comments and reflect that affect appropriately. Go to 5.29.

5.29   *The ability to reflect feeling is useful at most points in an interview and in response to a variety of statements. The following statement is made in the early stages of an interview.*

CLIENT: (a former drug addict discussing his rehabilitation) What can I do? I have no money, no skills, and no one to help me. I should do something, but I don't know what.

Choose the most appropriate response.

INTERVIEWER: You sound bewildered at the moment. (Go to 5.30)

INTERVIEWER: You sound as though you've given up hope. (Go to 5.31)

INTERVIEWER: All of that must be frightening to you. (Go to 5.32)

5.30   Your answer: *You sound bewildered at the moment.*
Correct. Your answer indicates that you've accurately identified the client's feelings. Proceed to 5.33.

5.31   Your answer: *You sound as though you've given up hope.*
This response indicates that you've focused on only part of what the client has said. By stating that he "should do something," he has shown that he has some desire to alter his situation. Return to 5.29 and try again.

5.32   Your answer: *All of that must be frightening to you.*
This response is an overstatement of the emotion expressed by the client. He has said that he has no money or skills and that he doesn't know what to do; however, he has not indicated that he's frightened about his situation. Your inaccurate reflection may puzzle or even irritate the client. Return to 5.29 and try again.

5.33   *It is appropriate to reflect clients' feelings at any time, regardless of the nature of those feelings (positive, negative, or ambivalent) and regardless of the direction of their expression (toward themselves, the interviewer, or the interviewing situation).*

CLIENT: There was a time when I felt depressed, but now, thanks to you, I don't feel that way anymore.

Choose the response that best reflects the client's feelings.

INTERVIEWER: We still have a long way to go. (Go to 5.34)

INTERVIEWER: You've come a long way, and you're thankful for that. (Go to 5.35)

INTERVIEWER: You feel that I've helped you a great deal, but it was you who did the work. (Go to 5.36)

5.34 Your answer: *We still have a long way to go.*
This response isn't a reflection of feeling: It's a piece of advice that is unlikely to benefit you or the client. Return to 5.33 and try again.

5.35 Your answer: *You've come a long way, and you're thankful for that.*
Correct. It's always appropriate to reflect feelings, even when they are directed toward you, the interviewer. Reflection of such feelings helps develop the interviewing relationship. Go to 5.37.

5.36 Your answer: *You feel that I've helped you a great deal, but it was you who did the work.*
This reflection of feeling detracts from the statement the client has made. This response denies the client the right to express his or her feelings. Return to 5.33 and try again.

5.37

Clients are occasionally out of touch with their feelings or are unable to discuss them. When their feelings are reflected back to them, however, they become more aware of them and are able to evaluate their appropriateness.

CLIENT: (describing the discovery of an accident in his home) I turned on the light and saw him lying on the floor. There was blood everywhere. It was unbelievable.

Choose the response that best reflects the client's feelings.

INTERVIEWER: What on earth did you do? (Go to 5.38)

INTERVIEWER: You must have been paralyzed. (Go to 5.39)

INTERVIEWER: You must have been shocked. (Go to 5.40)

5.38 Your answer: *What on earth did you do?*
This is an open question that will elicit factual information, but it will not lead to a discussion of feelings. At this point, your task is to help the client become aware of and accept the feelings that the incident aroused. Return to 5.37 and try again.

5.39 Your answer: *You must have been paralyzed.*
By using the word paralyzed, you infer a feeling that is not implied in the client's statement. Return to 5.37 and try again.

5.40	**Your answer:**	*You must have been shocked.* Correct. Your response allows the client to consider how he felt and indicates that discussing his feelings is appropriate. Go to 5.41.

5.41		*Reflecting feeling may be defined as responding to the underlying feeling or emotional aspect of a client's response while ignoring its cognitive aspects or content.*
	CLIENT:	(describing a coworker) You should see him. He's really something! Good looking! And I've got a date with him. **Choose the response that best reflects the client's feelings.**
	INTERVIEWER:	You're very excited about that! (Go to 5.42)
	INTERVIEWER:	Don't get your hopes up; it's only a date. (Go to 5.43)
	INTERVIEWER:	So you have a date with him. (Go to 5.44)

5.42	**Your answer:**	*You're very excited about that!* Correct. Your response ignores the cognitive aspects of what the client has said and reflects her feelings—her excitement. Go to 5.45.

5.43	**Your answer:**	*Don't get your hopes up; it's only a date.* This response is a piece of advice, not a reflection of feeling. You have ignored the feelings of the client and concentrated on the cognitive aspect of what she has said—her date and its meaning. Return to 5.41 and try again.

5.44	**Your answer:**	*So you have a date with him.* This response is a reflection of content—a repetition of a portion of what the client said. Although this response can be appropriate, it is not the response requested of you here. Return to 5.41 and try again.

5.45		*What the client says represents the content portion of a message. You should also be aware of how the client conveys a message. For example, the client may speak more quickly when communicating enthusiasm, more slowly when communicating discouragement, and so on.*
	CLIENT:	(with her head down, speaking in a low tone of voice) I was going to go skiing, but my mother is coming to stay with me. **Choose the response that best reflects the client's feelings.**
	INTERVIEWER:	It sounds as though that will be just as enjoyable. (Go to 5.46)
	INTERVIEWER:	You seem unhappy about that. (Go to 5.47)
	INTERVIEWER:	You must be looking forward to her arrival. (Go to 5.48)

5.46	**Your answer:**	*It sounds as though that will be just as enjoyable.* You haven't attended to how the client delivered her response. Her head was down and she spoke in a low tone of voice. It is unlikely that her behavior indicated that she was happy about what she was telling you. Return to 5.45 and try again.

5.47	**Your answer:**	*You seem unhappy about that.*

Correct. By noting her posture and tone of voice, you have accurately identified the client's feelings. Go to 5.49.

5.48	**Your answer:**	*You must be looking forward to her arrival.*

The client has made no mention of looking forward to her mother's arrival. This is your interpretation, not an accurate reflection of the client's feelings. Return to 5.45, and attend carefully to how the client delivers her message.

5.49

Discussing feelings is often an important prerequisite to solving problems. By selectively attending to and reflecting feelings, the interviewer helps the client talk about them.

CLIENT: I passed the test! I passed! I even got a good mark!

Choose the response that best reflects the client's feelings.

INTERVIEWER: You did it. You're just so excited. (Go to 5.50)

INTERVIEWER: You feel quite relieved to know you passed. (Go to 5.51)

INTERVIEWER: So, you got a good grade on your test. (Go to 5.52)

5.50	**Your answer:**	*You did it. You're just so excited.*

Correct. Your response focuses on how the client feels, not on what is being said. Go to 5.53.

5.51	**Your answer:**	*You feel quite relieved to know you passed.*

This response fails to reflect the client's exhilaration. Moreover, you've assumed a feeling of relief that may not be present. Return to 5.49 and try again.

5.52	**Your answer:**	*So, you got a good grade on your test.*

This response helps the client talk about grades and this particular test, but it doesn't help the discussion of feelings. You have ignored the emotional aspects of the response. Return to 5.49 and try again.

5.53

When you reflect feelings, you indicate that you can accurately sense the world as the client feels and perceives it; as a result, you build a good relationship with the client.

CLIENT: I'm glad I finally told you about it. You seem to understand. You seem to understand everything I tell you.

Choose the response that best reflects the client's feelings.

INTERVIEWER: You sound happy about that. (Go to 5.54)

INTERVIEWER: You should have told someone about this years ago. (Go to 5.55)

INTERVIEWER: It's a relief to be able to talk about it. (Go to 5.56)

| 5.54 | **Your answer:** | *You sound happy about that.* |
| | | This is a reflection of feeling, but a poor one. When you reflect feeling, it is important to identify the essence of the underlying feeling. Return to 5.53 and try again. |

| 5.55 | **Your answer:** | *You should have told someone about this years ago.* |
| | | This response ignores what the client has told you. Your advice is of no use to the client and will probably hinder the development of your relationship. Return to 5.53 and try again. |

| 5.56 | **Your answer:** | *It's a relief to be able to talk about it.* |
| | | Correct. You've indicated that you understand how the client feels now that he has shared this private information with you. Go to 5.57. |

5.57

An important aspect of reflecting feeling is the ability to summarize clearly the emotional aspects of clients' messages, enabling them to understand and deal with their own emotions.

CLIENT: (discussing her aging father) I'm so worried about him. He's working too hard. I don't know what to do to slow him down.

Choose the response that best reflects the client's feelings.

INTERVIEWER: You sound very worried about what your father will do to himself by working too hard. You want to help him, but you don't know what to do. You're frustrated by all of this—maybe even a little angry. (Go to 5.58)

INTERVIEWER: You're anxious, and you're concerned about what can be done for your father. (Go to 5.59)

INTERVIEWER: You sound worried. If I were you, I'd go and have a talk with the family doctor or maybe your minister. (Go to 5.60)

| 5.58 | **Your answer:** | *You sound very worried about what your father will do to himself by working too hard. You want to help him, but you don't know what to do. You're frustrated by all of this—maybe even a little angry.* |
| | | A reflection of feeling should be clear, concise, accurate. This reflection of feeling is confused, lengthy, and inaccurate. The client hasn't given any indication of frustration or anger. Return to 5.57 and try again. |

| 5.59 | **Your answer:** | *You're anxious and you're concerned about what can be done for your father.* |
| | | Correct. Your clear, concise summary will help the client identify and cope with her feelings. Go to 5.61. |

| 5.60 | **Your answer:** | *You sound worried. If I were you, I'd go and have a talk with the family doctor or maybe your minister.* |
| | | The impact of your reflection of feeling is diminished by the advice that followed it. Clients frequently work toward their own solutions when assisted by clear, concise reflections of feelings. Return to 5.57 and try again. |

5.61	Although a client's comments often refer to past, present, and future events, interviewers are most productive when they reflect and explore feelings that represent a client's current emotional state.

CLIENT: (separated from her husband 3 months ago) When he first told me that he was going to leave me, I was very angry, but now I'm managing by myself without his help.

Choose the response that best reflects the client's feelings.

INTERVIEWER: In the future, you'll be glad it worked out this way. (Go to 5.62)

INTERVIEWER: You were really angry when he first told you. (Go to 5.63)

INTERVIEWER: You're really proud that you have managed by yourself. (Go to 5.64)

5.62 **Your answer:** *In the future, you'll be glad it worked out this way.*
There is no basis for this assumption in what the client has told you (although it may be true). Moreover, reflections of feeling are most productive when they focus on a client's current emotional state. Return to 5.61 and try again.

5.63 **Your answer:** *You were really angry when he first told you.*
This response is a reflection of past feeling. You have attended to only one part of the client's response, not to her current feelings. Return to 5.61 and try again.

5.64 **Your answer:** *You're really proud that you have managed by yourself.*
Correct. You've accurately reflected the client's current feelings. This reflection will help her explore her feelings. Go to 5.65.

5.65	It is important to reflect and explore emotions that pertain to the immediate interview.

CLIENT: My father was a counselor. Boy, did I ever hate him. I'm sure all counselors are alike.

Choose the response that best reflects the client's feelings.

INTERVIEWER: I get the feeling you don't like counselors, including me. (Go to 5.66)

INTERVIEWER: You must have hated your father very much. (Go to 5.67)

INTERVIEWER: You shouldn't feel that all counselors are bad. (Go to 5.68)

5.66 **Your answer:** *I get the feeling you don't like counselors, including me.*
Correct. The client has implied that he hates all counselors, including you. With this response, you are asking him to explore his feelings about you. It is important that the client understand his feelings about you before he goes on to discuss other matters. Go to 5.69.

5.67 **Your answer:** *You must have hated your father very much.*
You've reflected the client's feelings, but you've missed the point of what he told you about counselors, of whom you are one. You must clarify the emotions the

client is expressing toward you before you can help him. Return to 5.65 and try again.

5.68   Your answer: *You shouldn't feel that all counselors are bad.*
This response denies the client the right to express, explore, and resolve his feelings. Return to 5.65 and try again.

5.69

Clients usually express mixed feelings, rather than single feelings, during interviews. One of the responsibilities of an interviewer is to help clients sort out their mixed feelings.

CLIENT: (discussing her estranged husband) I hope Joe will pay child support. He was no good, but he did provide. I worry about what we will do if he doesn't.

Choose the response that best reflects the client's feelings.

INTERVIEWER: It sounds as though you're depending on Joe to come through. (Go to 5.70)

INTERVIEWER: You hope that you'll have some financial support, but you're anxious because you're not sure that Joe will come through. (Go to 5.71)

INTERVIEWER: I guess you're anxious about being left without any financial support. (Go to 5.72)

5.70   Your answer: *It sounds as though you're depending on Joe to come through.*
This response reflects only one aspect of the client's feelings. She is displaying mixed emotions, and you should help her sort out these emotions. Return to 5.69 and try again.

5.71   Your answer: *You hope that you'll have some financial support, but you're anxious because you're not sure that Joe will come through.*
Correct. You've identified the client's mixed emotions, thereby helping her clarify her reaction and identify those feelings she wants to discuss further. Go to 5.73.

5.72   Your answer: *I guess you're anxious about being left without any financial support.*
You've reflected one aspect of the client's feelings, but not the mixed nature of these feelings. Help the client clarify the mixed state of her emotions and explore them. Return to 5.69 and try again.

5.73

Failure to identify and reflect a client's mixed emotions can result in undue attention to a single aspect of a problem to the exclusion of other equally important facets.

CLIENT: I didn't do well in the exams. I'm going home next week. I'm looking forward to seeing my parents, but I don't know what they will say about my grades.

Choose the response that best reflects the client's feelings.

	INTERVIEWER:	It's really going to be great for you to be home with your parents again. (Go to 5.74)
	INTERVIEWER:	You're quite anxious about your parents' reaction to your grades. (Go to 5.75)
	INTERVIEWER:	You're excited about seeing your parents again, but you're anxious about their reaction to your grades. (Go to 5.76)

5.74 **Your answer:** *It's really going to be great for you to be home with your parents again.*
You've responded to only one aspect of the mixed emotions the client has displayed. You should help the client sort out the mixed emotions being experienced. Return to 5.73 and try again.

5.75 **Your answer:** *You're quite anxious about your parents' reaction to your grades.*
The client has mixed feelings about the proposed trip home. You've responded to one of these feelings, but you haven't helped the client clarify her confused feelings. Return to 5.73 and try again.

5.76 **Your answer:** *You're excited about seeing your parents again, but you're anxious about their reaction to your grades.*
Correct. You've captured the mixed feelings the client has expressed. Go to 5.77.

5.77

Occasionally, a client may say one thing with words while saying something entirely different with actions.

CLIENT: (a widow talking about her only son, and crying) I'm really glad my son is going away to college.

Choose the response that best reflects the client's feelings.

INTERVIEWER: You must be extremely proud of him. Where is he going? (Go to 5.78)

INTERVIEWER: You're happy that he's going to college, but you'll be lonely without him. (Go to 5.79)

INTERVIEWER: You're very unhappy about his decision to go to college. (Go to 5.80)

5.78 **Your answer:** *You must be extremely proud of him. Where is he going?*
You've interrupted, rather than reflected, the client's feelings. Moreover, you've changed the topic. Return to 5.77 and try again.

5.79 **Your answer:** *You're happy that he's going to college, but you'll be lonely without him.*
Correct. Your statement reflects the feeling the client discussed as well as the emotion she displayed. Go to 5.81.

5.80 **Your answer:** *You're very unhappy about his decision to go to college.*
Although this reflection of feeling identifies the emotion associated with the client's nonverbal behavior, it fails to identify the feeling associated with her verbal message. Return to 5.77 and try again.

5.81

CLIENT: (fidgeting in his chair and wringing his hands) I'm glad I got in to see you today.

Choose the response that best reflects the client's feelings.

INTERVIEWER: You're pleased to be here, but you're somewhat anxious at the same time. (Go to 5.82)

INTERVIEWER: I'm glad you were able to see me. (Go to 5.83)

INTERVIEWER: You seem to be fairly anxious today. (Go to 5.84)

5.82 Your answer: *You're pleased to be here, but you're somewhat anxious at the same time.*
Correct. Your response reflects the mixed emotions the client is indicating at the moment. He is showing his anxiety by his nonverbal behavior while telling you that he is glad to see you. Go to 5.85.

5.83 Your answer: *I'm glad you were able to see me.*
Although your response reflects the verbal content of the client's communication, it ignores the message of his bodily movements—fidgeting and wringing his hands. Adequate reflection of feeling brings to a person's attention the mixed feelings that he or she communicates. Return to 5.81 and try again.

5.84 Your answer: *You seem to be fairly anxious today.*
You've accurately reflected the client's anxiety, but not his feelings about being glad to see you. Return to 5.81 and try again.

5.85

CLIENT: (smiling pleasantly) He doesn't know what the word share means. He believes that what's his is his and what's mine is his, too.

Choose the response that best reflects the client's feelings.

INTERVIEWER: And you don't agree with that, I suppose. (Go to 5.86)

INTERVIEWER: I'm puzzled. You're smiling, but you sound so resentful. (Go to 5.87)

INTERVIEWER: Because you're smiling, I get the feeling his attitude doesn't bother you very much. (Go to 5.88)

5.86 Your answer: *And you don't agree with that, I suppose.*
In making this response, you judge rather than reflect the client's feelings. Pay close attention to the client's verbal and nonverbal behavior, and select a response that reflects both. Return to 5.85 and try again.

5.87 Your answer: *I'm puzzled. You're smiling, but you sound so resentful.*
Correct. You've reflected the fact that the client is smiling while telling you of a situation he resents. By reflecting his verbal and nonverbal communications, you've asked him to clarify for both of you what he is feeling. Go to 5.89.

5.88   **Your answer:**   *Because you're smiling, I get the feeling his attitude doesn't bother you very much.* You've misunderstood either the client's smile or his comments. Your reflection is inaccurate and will probably irritate the client. Return to 5.85 and try again.

5.89

Reflection of feeling can help a client who is finding it difficult to continue in an interview.

CLIENT: (has been fidgeting for several minutes)

Choose the most appropriate response.

INTERVIEWER: (Turn to your desk and take notes until the client is ready to continue.) (Go to 5.90)

INTERVIEWER: Surely it can't be that difficult to talk about it; I can't help you until you do. (Go to 5.91)

INTERVIEWER: You seem very anxious about whatever it is you want to discuss with me. (Go to 5.92)

5.90   **Your answer:**   *(Turn to your desk and take notes until the client is ready to continue.)* Your response closes down all communication between you and the client, who is already finding it difficult to discuss his problem. Return to 5.89 and try again.

5.91   **Your answer:**   *Surely it can't be that difficult to talk about it; I can't help you until you do.* The client already realizes that you can't help him unless he tells you his story. However, when he realizes that you understand his difficulty in discussing his problem, he will be more likely to open up to you. Return to 5.89 and try again.

5.92   **Your answer:**   *You seem very anxious about whatever it is you want to discuss with me.* Correct. You've reflected the feeling the client is expressing by his behavior. This reflection will encourage him to relax and tell you his story. Go to 5.93.

5.93

Review Questions

It isn't necessary to use one phrase repeatedly when you reflect feelings; introductory phrases can provide some variation. Some of these phrases are *it seems that you feel . . . , you believe . . . , it sounds like . . . , in other words, you feel . . . , I gather that . . . , you really are . . . .* In your own experience, you will come across many other introductory phrases.

Write a reflection of feeling for each of the following statements. Then compare your responses with the examples that follow.

1. I'm just fed up with my father. He's always telling me what to do.
2. The future looks very good; I'm looking forward to it.
3. I can't stand those long lines in the bank.
4. Each time I go to the dentist, I tremble.
5. It's going to be so great—I can't wait to get started.
6. There's no future, so why should I do anything?
7. By the time he gets home, I'm just so concerned.
8. I just lost my job because of those politicians and their economic mismanagement.

The following reflections of feeling could be used in response to the preceding statements. Note the varied introductory phrases and the range of emotions.

1. You sound irritated (angry, hostile).
2. You're really hopeful (happy, optimistic) about your future.
3. Waiting makes you feel very impatient (fed up, annoyed).
4. You feel very tense (anxious, perturbed) on those occasions.
5. It sounds as though you're full of enthusiasm (excitement, eagerness).
6. You seem to be depressed (down, discouraged) right now.
7. In other words, you become anxious (tense, uptight) when he is delayed.
8. You're really bitter (resentful, frustrated) about what's happened to you.

If you had difficulty producing appropriate reflections of feeling for three or more of these situations, you should work through this chapter again.

Points to Remember about Reflecting Feeling

1. WHEN IDENTIFYING FEELINGS:
   a. Attend to the affective component of what the client says.
   b. Attend to the client's behavior (posture, voice tone, rate of delivery, and other mannerisms).
   c. Use a wide range of words to label emotions appropriately.
   d. Identify the full range of client emotions.
2. WHEN REFLECTING FEELINGS:
   a. Use an appropriate introductory phrase followed by a clear and concise summary of the feelings the client seems to be experiencing.
   b. Reflect mixed emotions.
   c. Don't repeat the client's exact words.
   d. Focus on current feelings, using the present tense.
   e. Use a wide range of introductory phrases, not just a few overworked ones.

3. REFLECTION OF FEELING:
   a. Helps clients become aware of their feelings.
   b. Helps clients accept and explore their feelings.
   c. Helps you demonstrate that you understand how the client is experiencing the world.
   d. Helps develop a strong relationship between you and the client.
4. REFLECT FEELING:
   a. In response to all types of emotion (positive, negative, or ambivalent).
   b. Regardless of the direction of the emotion (toward the client, the interviewer, or others).
   c. When the client has feelings about the interview that may impede its progress.
   d. When the client is having difficulty telling his or her story.

5.1

Activity Unit

Find two others who have also read this chapter. Role play brief interviews (no more than five interviewer responses) alternating the client, interviewer, and observer roles. Record each brief interview on audio- or videotape.

Interviewer role Practice reflections of feeling in response to the client.

Client role Be cooperative and provide information having an affective component. Clients may want to role play the problem of a friend or family member or to enact one of the following roles:

1. A parent whose child has just been taken into custody by the local authorities.
2. A homeowner whose mortgage has just been foreclosed for failure to make payments.
3. An individual who is happy about getting married but has some last-minute concerns.
4. A person who is dissatisfied with his or her job but is unable to find other employment.

Observer role Operate the equipment and complete the Practice Interview Checklist (see Table 5.1).

Postinterview discussion Using the taped interview and the Practice Interview Checklist data, the participants should discuss the performance of the interviewer. All participants can learn from this discussion, which should be nonjudgmental, focus on positive as well as less effective responses, and stimulate improved performance for all.

Table 5.1

Practice Interview Checklist

Activity	Interviewer response[a]				
	1	2	3	4	5
On topic					
Reflection of feeling					
Effective response[b]					

[a] Check the activities that characterize each response.

[b] Indicate the degree of effectiveness on a scale from 1 to 5 in which 1 is ineffective and 5 is very effective.

5.2

Activity Unit

Using the tape of the interview produced in Activity Unit 5.1 or the tape of another practice interview, enter the original interview responses below. Then formulate a better response to replace each original one in the space provided.

Original response 1 _____

_____

Better response _____

_____

_____

Original response 2 _____

_____

Better response _____

_____

_____

Original response 3 _____

_____

Better response _____

_____

_____

Original response 4 _____

_____

_____

Better response _____

_____

_____

Original response 5 _____

_____

_____

Better response _____

_____

_____

Cultural Considerations

1. To use reflections of feeling with clients of other cultural affiliations, the interviewer needs to know how emotions are expressed within the culture.
2. A client's cultural affiliation determines the degree of response to a reflection of feeling. In some cultures, for example, Asian and Asian American, restraint of feelings is valued, and as a result, little or no response to a reflection of feeling is given. Further, Native North Americans, Latino men, African American men, and European American men find exploring emotions difficult.
3. Direct eye contact is avoided in some cultures. With clients from cultures in which this is the case, the interviewer should be cautious in interpreting lack of eye contact as a sign of shyness, embarrassment, discomfort, anxiety, or depression.
4. There is evidence that facial expressions of emotions are not culture bound. However, the client's culture determines how and when facial expressions of emotions occur.

InfoTrac® College Edition Keyword Search Terms

Empathy AND Counseling

Chapter Six

Developing an Individual Style

C hapters 2 through 5 each focused on single skills. This chapter is intended to help you integrate these skills. After completing this chapter, you should be able to:

1. Attend to the essence of a client's story during an interview.
2. Integrate the use of open questions, minimal encouragement, paraphrasing, summarization, and reflection of feeling throughout an interview.
3. Decide which skill is appropriate for you to use in response to a client.

In the first five chapters of this book, you learned the skills of attending behavior, effective questioning, minimal encouragement, paraphrasing, summarizing, and reflection of feeling. The method of training followed in these chapters is the *single-skills approach*. This kind of training has been used to teach individuals a number of leisure activities, such as baseball, softball, skiing, and bowling, each of which consists of a number of single skills. For example, a softball player has to learn stance, grip, wrist action, leg action, head position, and so on when batting. When players concentrate on only one skill at a time, their batting remains imperfect. However, after they have mastered a number of skills and are able to integrate them, their batting improves dramatically. Just as beginning batters reach a point at which it is necessary to integrate single skills into a fluid, cohesive whole, beginning interviewers must also reach a point of integration.

When a beginning interviewer first starts to integrate the basic skills, it can feel strange and awkward because there seems to be so much to master and remember. However, once you begin to integrate the introductory skills of attending, questioning, paraphrasing, summarization, and reflection of feeling, the process becomes more comfortable. With practice and comfort in interviewing comes the ability to identify when a particular skill is needed to advance the interview. This leads to the development of your own particular style of interviewing. Remember, the foundation for effective interviewing is the ability to listen to and understand the client.

This chapter is divided into two sections. The first section helps you review and consolidate the interviewing skills learned in previous chapters. The second section of the chapter helps you integrate these specific skills in an interview.

The program in this chapter focuses on an interview between a man and a social worker (male or female) in a community agency. The man has requested the interview, saying only that the matter is urgent. Go to 6.1.

6.1	*Attending behavior—the basic interviewing skill—consists of appropriate eye contact, a relaxed, professional body posture, and verbal responses that indicate you understand the client's message.*
INTERVIEWER:	What would you like to talk about today?
CLIENT:	Well, it's my son. I just don't know what to do with him.
	Choose the most appropriate response.
INTERVIEWER:	How old is your son? (Go to 6.2)
INTERVIEWER:	(after a brief silence) Your son? (Go to 6.3)
INTERVIEWER:	How does your son feel about you? (Go to 6.4)

6.2   Your answer: *How old is your son?*
This question may be appropriate later in the interview, but it's inappropriate at this point. Attend to what the client is communicating and respond in a way that indicates you are following him. Return to 6.1 and try again.

6.3   Your answer: *(after a brief silence) Your son?*
Correct. Your brief silence invited the client to continue. However, when you sensed that he wasn't going to continue, you repeated a portion of his statement. This minimal encouragement indicates that you are following his comments. Go to 6.5.

6.4	**Your answer:**	*How does your son feel about you?*

You've changed the topic. When you attend to the client's communication, you help the client tell you his story in his own way. Return to 6.1 and try again.

6.5

In the early stages of working with a client, it's important to listen; problem solving and information giving come much later.

CLIENT: Yes, my son. Last year he was unmanageable. And just recently he's been shoplifting.

Choose the most appropriate response.

INTERVIEWER: He's been shoplifting . . . (Go to 6.6)

INTERVIEWER: Were you able to put a stop to that? (Go to 6.7)

INTERVIEWER: Well, I'm sure we can deal with that. (Go to 6.8)

6.6	**Your answer:**	*He's been shoplifting . . .*

Correct. You've indicated that you're following the client and are interested in what he's saying. Go to 6.9.

6.7	**Your answer:**	*Were you able to put a stop to that?*

This irrelevant question, with its implied advice, may irritate or anger the client. Return to 6.5 and try again.

6.8	**Your answer:**	*Well, I'm sure we can deal with that.*

This may be true, but the son is not the client, his father is. Your task is to attend to what the client is saying and help him explore and clarify his problem. After you've done this, you can help him identify the best way to handle the problem. Return to 6.5 and try again.

6.9

Your attention or lack of attention, as indicated by your responses and behavior, significantly influences your relationship with a client.

CLIENT: Yes. Last week the police were around with a warrant. And it damn well wasn't for one offense but 14!

Choose the most appropriate response.

INTERVIEWER: Did they arrest him? (Go to 6.10)

INTERVIEWER: Is he sorry about what he's done? (Go to 6.11)

INTERVIEWER: You sound angry about that. (Go to 6.12)

6.10	**Your answer:**	*Did they arrest him?*

Although you may need this information, this is not the time to request it. You've indicated that your thoughts are elsewhere and that you aren't focusing on what the client is telling you. Return to 6.9 and try again.

6.11   **Your answer:**   *Is he sorry about what he's done?*
You've changed the topic to one that may be of interest to you but may not be important to the client at this time. Return to 6.9 and try again.

6.12   **Your answer:**   *You sound angry about that.*
Correct. The client has expressed considerable anger. By reflecting this anger, you've indicated that you're attending to what he is saying as well as to how he feels about what he is saying. Go to 6.13.

6.13   *Open questions encourage a client to discuss problems with you.*

CLIENT: Yes, I am, because that means he's been doing it for some time. That, along with his other behavior, would make anyone angry.

Choose the most appropriate response.

INTERVIEWER: Is your son on drugs? (Go to 6.14)

INTERVIEWER: Can you expand on that for me? (Go to 6.15)

INTERVIEWER: Do you think your son realizes that you're angry? (Go to 6.16)

6.14   **Your answer:**   *Is your son on drugs?*
This closed question invites a "yes" or "no" answer and changes the topic. By using a closed question, you assume responsibility for continuing the interview. An open question, on the other hand, helps the client explore his problem and gives him an opportunity to continue the discussion. Return to 6.13 and try again.

6.15   **Your answer:**   *Can you expand on that for me?*
Correct. This open question invites the client to tell you more about his problems with his son. Go to 6.17.

6.16   **Your answer:**   *Do you think your son realizes that you're angry?*
This closed question requires only a "yes" or "no" answer. The client can now sit back and wait for you to talk again. An open question, on the other hand, could help the client discuss the problem concerning his son. Return to 6.13 and try again.

6.17   *Open questions often begin with what, how, or could. What often elicits factual information; how is people oriented; could encourages a client to continue. Although why can be used effectively, its use often places a client on the defensive.*

CLIENT: Well, for about a year now, we've had trouble with him. He doesn't come home for meals, and he gets in late at night.

Choose the most appropriate response.

INTERVIEWER: How have you handled his behavior? (Go to 6.18)

| INTERVIEWER: | Does he ever stay out all night? (Go to 6.19) |
| INTERVIEWER: | Why do you let him get away with it? (Go to 6.20) |

6.18 Your answer: *How have you handled his behavior?*
Correct. This open question invites the client to continue telling you about his son's behavior and how it has affected him. Go to 6.21.

6.19 Your answer: *Does he ever stay out all night?*
This closed question may require several follow-up questions. An open question, on the other hand, would elicit more information and encourage the client to assume responsibility for the interview's progress. Return to 6.17 and try again.

6.20 Your answer: *Why do you let him get away with it?*
You've assumed that the client hasn't tried to change his son's behavior. This assumption may make the client feel defensive, especially if he has tried to make his son come home at night. Return to 6.17 and try again.

6.21

Reflection of feeling focuses on the emotional aspects of a client's comments.

CLIENT: We ignored his lateness at first because it happened only occasionally. But lately he's been out every night after midnight, and sometimes after one or two o'clock. I stay up and give him a piece of my mind.

Choose the most appropriate response.

INTERVIEWER:	When do you think he should be in? (Go to 6.22)
INTERVIEWER:	How does your wife feel about all of this? (Go to 6.23)
INTERVIEWER:	You really feel angry when he comes in late. (Go to 6.24)

6.22 Your answer: *When do you think he should be in?*
This closed question changes the topic. The client was telling you how he feels about his son coming in late. Return to 6.21 and try again.

6.23 Your answer: *How does your wife feel about all of this?*
You've changed the topic by asking a question that has nothing to do with the topic being discussed. Return to 6.21 and try again.

6.24 Your answer: *You really feel angry when he comes in late.*
Correct. The client has been telling you how he reacts when his son comes in late, and you've reflected the feeling implied by his reaction. Go to 6.25.

6.25

Correctly identify a client's feelings before you attempt to reflect them.

CLIENT: Yes, I feel angry when it happens, but the next morning, I just feel upset about it all. After all, it's his life he's messing up. I wish I could just make him understand that we love him.

Choose the most appropriate response.

INTERVIEWER: So your feelings are mixed. You feel anger, concern, and love. (Go to 6.26)

INTERVIEWER: You don't feel angry the next morning. (Go to 6.27)

INTERVIEWER: Do you really love him? (Go to 6.28)

6.26   Your answer: *So your feelings are mixed. You feel anger, concern, and love.*
Correct. The client has displayed a number of feelings, which you've identified and reflected. By doing so, you've given him a chance to identify his feelings and clarify them. Go to 6.29.

6.27   Your answer: *You don't feel angry the next morning.*
Your response focuses on only one aspect of the client's statement. It's important to him to identify and explore all of the feelings he is exhibiting. Return to 6.25 and try again.

6.28   Your answer: *Do you really love him?*
This closed question focuses on only one aspect of the client's statement. You haven't encouraged him to explore his mixed feelings. Return to 6.25 and try again.

6.29   *Summarization allows a client to review what has been said in an interview.*

CLIENT: Yes, I guess I'm concerned for him and for us. I don't know how we can get him to change.

Choose the most appropriate response.

INTERVIEWER: Maybe you should encourage him to stay home. (Go to 6.30)

INTERVIEWER: So, he's in trouble with the law, he doesn't come home at night, and you want to find a solution for these problems. (Go to 6.31)

INTERVIEWER: Well, the courts will probably decide that for you. (Go to 6.32)

6.30   Your answer: *Maybe you should encourage him to stay home.*
Your response indicates that you don't think the client has tried to do this. If he has, your interpretation could make him feel resentful. If he hasn't found a way to encourage his son to stay home, your comment may make him feel defensive and cause him to withdraw from his interaction with you. Return to 6.29 and try again.

6.31   Your answer: *So, he's in trouble with the law, he doesn't come home at night, and you want to find a solution for these problems.*
Correct. This is a good summary of what the client has told you, and it focuses his attention on what he has discussed with you up to this point. Go to 6.33.

6.32	**Your answer:**	*Well, the courts will probably decide that for you.*

You've demonstrated your lack of attention to the client's comments and your lack of interest in how he feels. Your reply doesn't help him clarify his problem. Return to 6.29 and try again.

6.33

Paraphrasing the content of statements made during an interview is another way to focus a client's attention on what he or she has told you.

CLIENT: That's about it. My wife and I have tried just about everything, but nothing seems to work.

Choose the most appropriate response.

INTERVIEWER: Perhaps you should bring your wife in to see me. (Go to 6.34)

INTERVIEWER: Do you think there is a solution you may have overlooked? (Go to 6.35)

INTERVIEWER: Everything you've thought of has failed. (Go to 6.36)

6.34   **Your answer:**   *Perhaps you should bring your wife in to see me.*
Your response indicates that you weren't following the client attentively. Return to 6.33 and try again.

6.35   **Your answer:**   *Do you think there is a solution you may have overlooked?*
This closed question calls for a "yes" or "no" answer. Return to 6.33 and try again.

6.36   **Your answer:**   *Everything you've thought of has failed.*
Correct. This restatement of what the client has told you focuses on the problem he is discussing. Go to 6.37.

6.37

Minimal encouragements indicate that you are attending to the client's story and invite a client to continue a discussion.

CLIENT: Yes. Did I tell you he was adopted?

Choose the most appropriate response.

INTERVIEWER: From where? (Go to 6.38)

INTERVIEWER: Hmmm . . . (Go to 6.39)

INTERVIEWER: Uhm-hmm . . . go on. (Go to 6.40)

6.38   **Your answer:**   *From where?*
This response interrupts the client's discussion. Return to 6.37 and try again.

6.39   **Your answer:**   *Hmmm . . .*
This response suggests uncertainty. Minimal encouragements should facilitate discussion, not confuse it. Return to 6.37 and try again.

6.40	**Your answer:**	*Uhm-hmm . . . go on.* Correct. This minimal encouragement invites the client to continue the discussion with you.

The remaining frames of this chapter give you an opportunity to integrate the interviewing skills discussed thus far. A number of alternative routes through the program are available. As a result, you can advance a number of frames, depending on the alternative you choose. After you've completed the program, return to 6.41 and follow an alternative route. By working through this section twice, you will see that different combinations of skills and responses can facilitate progress during an interview. Go to 6.41.

6.41

Integrating interviewing skills develops your own natural mode of interviewing; that is, you use each skill when you feel it is appropriate.

CLIENT: We adopted him when he was 6 months old because we couldn't have children of our own.

Choose an appropriate response.

INTERVIEWER: Was he a good baby? (Go to 6.42)

INTERVIEWER: Could you tell me more about that? (Go to 6.43)

INTERVIEWER: You really wanted a family. (Go to 6.44)

6.42	**Your answer:**	*Was he a good baby?* Your response requires only a "yes" or "no" answer, and it doesn't help the client continue the discussion. Moreover, you've indicated that you're interested in what the baby was like, not in what the client is telling you. Return to 6.41 and try again.

6.43	**Your answer:**	*Could you tell me more about that?* Correct. This open question gives the client the responsibility for continuing the discussion. Go to 6.45.

6.44	**Your answer:**	*You really wanted a family.* Correct. This reflection of feeling encourages the client to tell you more about his wish to have a family. Go to 6.46.

6.45

CLIENT: Well, my wife and I had been married for 6 years, and we didn't have any children. We really wanted to have a family.

Choose an appropriate response.

INTERVIEWER: Having a family was important to you. (Go to 6.50)

INTERVIEWER: Did you see a doctor about your problem? (Go to 6.51)

INTERVIEWER: You felt very concerned because you had no children. (Go to 6.52)

6.46

CLIENT: Well, yes. For 6 years we'd been trying to have children.

Choose an appropriate response.

INTERVIEWER: Could you tell me more about that? (Go to 6.47)

INTERVIEWER: Does anyone else in your family have this problem? (Go to 6.48)

INTERVIEWER: But without success. (Go to 6.49)

6.47 **Your answer:** *Could you tell me more about that?*
Correct. This open question encourages the client to continue. Go to 6.54.

6.48 **Your answer:** *Does anyone else in your family have this problem?*
This response won't help the client deal with his current problem. Return to 6.46 and try again.

6.49 **Your answer:** *But without success.*
Correct. This minimal encouragement indicates that you're attending to the client's story and invites him to continue. Go to 6.53.

6.50 **Your answer:** *Having a family was important to you.*
Correct. This restatement of content indicates that you are attending to the discussion, and it invites the client to continue. Go to 6.56.

6.51 **Your answer:** *Did you see a doctor about your problem?*
This closed question doesn't help the client expand on what he is telling you. As soon as he answers "yes" or "no," you have to assume responsibility for the next response. Return to 6.45 and try again.

6.52 **Your answer:** *You felt very concerned because you had no children.*
Correct. This reflection focuses on the client's feelings about having a family. Go to 6.55.

6.53

CLIENT: Right. So, finally we decided to see our doctor about our problem. That was a hard decision to make.

Choose an appropriate response.

INTERVIEWER: How did you feel after you'd seen your doctor? (Go to 6.66)

INTERVIEWER: Deciding to see your doctor was difficult. (Go to 6.67)

INTERVIEWER: The doctor reassured you and your wife. (Go to 6.68)

6.54   CLIENT: My wife and I both had checkups, and there was no apparent reason for our inability to have children. So, we decided to do something about it . . . and we did.

Choose an appropriate response.

INTERVIEWER: You finally came to a decision. (Go to 6.63)

INTERVIEWER: You decided . . . (Go to 6.64)

INTERVIEWER: Did you consider becoming foster parents? (Go to 6.65)

6.55   CLIENT: Yes. We really wanted children. Also, people began to leave us out of parties and things because they had families and we didn't.

Choose an appropriate response.

INTERVIEWER: So, it was a social concern as well as a personal problem. (Go to 6.60)

INTERVIEWER: Go on. (Go to 6.61)

INTERVIEWER: Your social life was all-important to you. (Go to 6.62)

6.56   CLIENT: I guess so. All of our friends had families, and everyone kept asking us when we were going to start ours. We were so uncomfortable.

Choose an appropriate response..

INTERVIEWER: Maybe you should have told them that it wasn't their concern. (Go to 6.57)

INTERVIEWER: It became really embarrassing for you. (Go to 6.58)

INTERVIEWER: And then what happened? (Go to 6.59)

6.57   **Your answer:**   *Maybe you should have told them that it wasn't their concern.*
Your response indicates that you've allowed your own feelings on the subject to interfere with your attentive listening. Clients profit from working on their own solutions to their problems. Return to 6.56 and try again.

6.58   **Your answer:**   *It became really embarrassing for you.*
Correct. This reflection focuses attention on the client's feelings and invites him to continue. Go to 6.69.

6.59   **Your answer:**   *And then what happened?*
Correct. This brief open question indicates that you are attending to the client, and it encourages him to expand on what he is telling you. Go to 6.70.

6.60   **Your answer:**   *So, it was a social concern as well as a personal problem.*
Correct. This restatement of content indicates that you are following the client and gives him a point of departure for his next comments. Go to 6.71.

6.61	**Your answer:**	*Go on.*
		Correct. This minimal encouragement indicates that you are following the client's story and would like him to continue. Go to 6.72.

6.62	**Your answer:**	*Your social life was all-important to you.*
		This leading statement will probably irritate the client. You can help the client explore his problem by attending carefully to what he is saying. Return to 6.55 and try again.

6.63	**Your answer:**	*You finally came to a decision.*
		Correct. This restatement of content indicates that you are attending to the client's comments and encourages him to continue the discussion with you. Go to 6.73.

6.64	**Your answer:**	*You decided . . .*
		Correct. This minimal encouragement indicates that you are following the client and invites him to continue. Go to 6.74.

6.65	**Your answer:**	*Did you consider becoming foster parents?*
		You've changed the topic of the discussion. Return to 6.54 and try again.

6.66	**Your answer:**	*How did you feel after you'd seen your doctor?*
		Correct. This open question focuses on the client's comments and asks him to continue the discussion. Go to 6.75.

6.67	**Your answer:**	*Deciding to see your doctor was difficult.*
		Correct. Your response accurately reflects the content of the client's statement and indicates that you're attending to what he is saying. At the same time, the information elicited by this response will enable you to check your understanding of the client's message. Go to 6.76.

6.68	**Your answer:**	*The doctor reassured you and your wife.*
		Your reflection is an inaccurate one because you have no information on which to base it. The doctor may have confirmed their worst nightmares and made them feel very depressed. Return to 6.53 and try again.

6.69

	CLIENT:	Yes. We didn't even go out much because we were afraid that the topic would come up. We had to do something.
		Choose an appropriate response..
	INTERVIEWER:	You were ashamed, so you just hid. (Go to 6.98)
	INTERVIEWER:	Uhm-hmm. (Go to 6.99)
	INTERVIEWER:	What did you decide to do? (Go to 6.100)

6.70   CLIENT: Well, we went to see a doctor, and he said that we would probably never have children of our own. So, after talking with him, we looked into adopting a child.

Choose an appropriate response.

INTERVIEWER: And then . . . (Go to 6.95)

INTERVIEWER: That was one alternative that offered some hope. (Go to 6.96)

INTERVIEWER: Did you arrange the adoption through your family doctor? (Go to 6.97)

6.71   CLIENT: Yes. We felt that we had to do something.

Choose an appropriate response.

INTERVIEWER: I'm not sure you had to feel that way. (Go to 6.92)

INTERVIEWER: And so . . . (Go to 6.93)

INTERVIEWER: What did you decide to do? (Go to 6.94)

6.72   CLIENT: I guess they didn't want to embarrass us by talking about their children in front of us. That wouldn't happen if they just didn't invite us.

Choose an appropriate response.

INTERVIEWER: So you were left out of things. (Go to 6.89)

INTERVIEWER: I sense that their behavior hurt you. (Go to 6.90)

INTERVIEWER: I guess there were other things you could have done. (Go to 6.91)

6.73   CLIENT: Yes. It was a difficult decision to make, but we decided to adopt a child.

Choose an appropriate response.

INTERVIEWER: A big decision . . . (Go to 6.86)

INTERVIEWER: How do you feel about that decision now? (Go to 6.87)

INTERVIEWER: How did your friends react? (Go to 6.88)

6.74   CLIENT: Yes. We decided to adopt. We had to wait while they checked us out, but we made it. What a day that was!

Choose an appropriate response.

INTERVIEWER: What kinds of things did they check? (Go to 6.83)

INTERVIEWER: So you were able to adopt. (Go to 6.84)

INTERVIEWER: How exciting that must have been! (Go to 6.85)

6.75

CLIENT: Well, we were told that we could have children, but it would be difficult. The doctor suggested that we adopt a child.

Choose an appropriate response.

INTERVIEWER: So, it seemed that adoption was the answer. (Go to 6.80)

INTERVIEWER: Did you reject that idea at first? (Go to 6.81)

INTERVIEWER: So, then . . . (Go to 6.82)

6.76

CLIENT: It sure was, but after the doctor told us, we were glad that we'd gone.

Choose an appropriate response.

INTERVIEWER: What did the doctor tell you? (Go to 6.77)

INTERVIEWER: You really felt relieved then. (Go to 6.78)

INTERVIEWER: It must have been good news. (Go to 6.79)

6.77   **Your answer:**   *What did the doctor tell you?*
Correct. This open question is on topic.
 Let's review your last few responses. In working your way from 6.41, you've reflected feeling, used a minimal encouragement, reflected content, and asked an open question. By now, you're aware that a variety of responses are appropriate at most points in an interview. Now return to 6.41 and choose another appropriate response. You've already chosen 6.44, so don't choose it again. If this is your second time through, go to 6.101.

6.78   **Your answer:**   *You really felt relieved then.*
Correct. This is an appropriate reflection of feeling.
 Let's review your last few responses. In working your way from 6.41, you've reflected feeling, used a minimal encouragement, reflected content, and reflected feeling again. By now, you're aware that a variety of responses are appropriate at most points in an interview. Now return to 6.41 and choose another appropriate response. You've already chosen 6.44, so don't choose it again. If this is your second time through, go to 6.101.

6.79   **Your answer:**   *It must have been good news.*
The client's previous statement doesn't suggest that the news he received was good; therefore, your statement is an assumption that may be incorrect. Return to 6.76 and try again.

6.80   **Your answer:**   *So, it seemed that adoption was the answer.*
Correct. This is an accurate paraphrase.
 Let's review your last few responses. In working your way from 6.41, you've reflected feeling, used a minimal encouragement, asked an open question, and reflected content. By now, you're aware that a variety of responses

are appropriate at most points in an interview. Now return to 6.41 and choose another appropriate response. You've already chosen 6.44, so don't choose it again. If this is your second time through, go to 6.101.

6.81   Your answer: *Did you reject that idea at first?*
This question may irritate the client and hinder your interaction with him. Return to 6.75 and try again.

6.82   Your answer: *So, then . . .*
Correct. This minimal encouragement will help the client continue.

Let's review your last few responses. In working your way from 6.41, you've reflected feeling, used a minimal encouragement, asked an open question, and used another minimal encouragement. By now, you're aware that a variety of responses are appropriate at most points in an interview. Now return to 6.41 and choose another appropriate response. You've already chosen 6.44, so don't choose it again. If this is your second time through, go to 6.101.

6.83   Your answer: *What kinds of things did they check?*
This response is irrelevant to the discussion. Return to 6.74 and try again.

6.84   Your answer: *So you were able to adopt.*
Correct. This is an appropriate reflection of content.

Let's review your last few responses. In working your way from 6.41, you've reflected feeling, asked an open question, used a minimal encouragement, and reflected content. By now, you're aware that a variety of responses are appropriate at most points in an interview. Now return to 6.41 and choose another appropriate response. You've already chosen 6.44, so don't choose it again. If this is your second time through, go to 6.101.

6.85   Your answer: *How exciting that must have been!*
Correct. This is an appropriate reflection of feeling.

Let's review your last few responses. In working your way from 6.41, you've reflected feeling, asked an open question, used a minimal encouragement, and reflected feeling again. By now, you're aware that a variety of responses are appropriate at most points in an interview. Now return to 6.41 and choose another appropriate response. You've already chosen 6.44, so don't choose it again. If this is your second time through, go to 6.101.

6.86   Your answer: *A big decision . . .*
Correct. This minimal encouragement will help the client continue the discussion.

Let's review your last few responses. In working your way from 6.41, you've reflected feeling, asked an open question, reflected content, and used a minimal encouragement. By now, you're aware that a variety of responses are appropriate at most points in an interview. Now return to 6.41 and choose another appropriate response. You've already chosen 6.44, so don't choose it again. If this is your second time through, go to 6.101.

6.87 **Your answer:** *How do you feel about that decision now?*
Correct. This open question is on topic.
   Let's review your last few responses. In working your way from 6.41, you've reflected feeling, used an open question, reflected content, and used an open question again. By now, you're aware that a variety of responses are appropriate at most points in an interview. Now return to 6.41 and choose another appropriate response. You've already chosen 6.44, so don't choose it again. If this is your second time through, go to 6.101.

6.88 **Your answer:** *How did your friends react?*
You've changed the topic, indicating to your client that you are more interested in pursuing your own thoughts than in attending to what he was discussing. Return to 6.73 and try again.

6.89 **Your answer:** *So you were left out of things.*
Correct. This is an appropriate reflection of content.
   Let's review your last few responses. In working your way from 6.41, you've used an open question, reflected feeling, used a minimal encouragement, and reflected content. By now, you're aware that a variety of responses are appropriate at most points in an interview. Now return to 6.41 and choose another appropriate response. You've already chosen 6.43, so don't choose it again. If this is your second time through, go to 6.101.

6.90 **Your answer:** *I sense that their behavior hurt you.*
Correct. This is an accurate reflection of feeling.
   Let's review your last few responses. In working your way from 6.41, you've used an open question, reflected feeling, used a minimal encouragement, and reflected feeling again. By now, you're aware that a variety of responses are appropriate at most points in an interview. Now return to 6.41 and choose another appropriate response. You've already chosen 6.43, so don't choose it again. If this is your second time through, go to 6.101.

6.91 **Your answer:** *I guess there were other things you could have done.*
The client has just indicated that he and his wife felt left out. Your response indicates that you don't understand how he feels. You've interpreted his statement from your own point of view. Return to 6.72 and try again.

6.92 **Your answer:** *I'm not sure you had to feel that way.*
This statement denies the client the right to his own feelings, preventing him from resolving any difficulties his feelings may have caused. Return to 6.71 and try again.

6.93   Your answer: *And so . . .*

Correct. This minimal encouragement will help the client continue.

Let's review your last few responses. In working your way from 6.41, you've used an open question, reflected feeling, reflected content, and used a minimal encouragement. By now, you're aware that a variety of responses are appropriate at most points in an interview. Now return to 6.41 and choose another appropriate response. You've already chosen 6.43, so don't choose it again. If this is your second time through, go to 6.101.

6.94   Your answer: *What did you decide to do?*

Correct. This open question focuses on the topic the client is discussing.

Let's review your last few responses. In working your way from 6.41, you've used an open question, reflected feeling, reflected content, and used an open question again. By now, you're aware that a variety of responses are appropriate at most points in an interview. Now return to 6.41 and choose another appropriate response. You've already chosen 6.43, so don't choose it again. If this is your second time through, go to 6.101.

6.95   Your answer: *And then . . .*

Correct. This minimal encouragement will help the client continue.

Let's review your last few responses. In working your way from 6.41, you've used an open question, reflected content, used another open question, and used a minimal encouragement. By now, you're aware that a variety of responses are appropriate at most points in an interview. Now return to 6.41 and choose another appropriate response. You've already chosen 6.43, so don't choose it again. If this is your second time through, go to 6.101.

6.96   Your answer: *That was one alternative that offered some hope.*

Correct. This is an accurate reflection of feeling.

Let's review your last few responses. In working your way from 6.41, you've used an open question, reflected content, used another open question, and reflected feeling. By now, you're aware that a variety of responses are appropriate at most points in an interview. Now return to 6.41 and choose another appropriate response. You've already chosen 6.43, so don't choose it again. If this is your second time through, go to 6.101.

6.97   Your answer: *Did you arrange the adoption through your family doctor?*

This closed question changes the topic. If this information is important to the client, he will convey it to you in his own time. Return to 6.70 and try again.

6.98   Your answer: *You were ashamed, so you just hid.*

This reflection of feeling is excessive. Don't overinterpret or make assumptions about the client's comments. Return to 6.69 and try again.

6.99   Your answer: *Uhm-hmm.*

Correct. This minimal encouragement invites the client to continue the discussion.

Let's review your last few responses. In working your way from 6.41, you've used an open question, reflected content, reflected feeling, and used a minimal encouragement. By now, you're aware that a variety of responses are appropriate at most points in an interview. Now return to 6.41 and choose another appropriate response. You've already chosen 6.43, so don't choose it again. If this is your second time through, go to 6.101.

6.100 Your answer: *What did you decide to do?*

Correct. This open question indicates that you're following and invites the client to continue.

Let's review your last few responses. In working your way from 6.41, you've used an open question, reflected content, reflected feeling, and used another open question. By now, you're aware that a variety of responses are appropriate at most points in an interview. Now return to 6.41 and choose another appropriate response. You've already chosen 6.43, so don't choose it again. If this is your second time through, go to 6.101

6.101

Review Questions

This chapter has been concerned with the consolidation and integration of specific interviewing skills. In an actual interview, the distinction between skills isn't always apparent. At times, it's appropriate to employ several skills in one statement: for example, "Uhm-hmm. . . . You sound angry" (minimal encouragement, reflection of feeling). Or "You sound concerned. Could you tell me more about that?" (reflection of feeling, open question). Your task is to integrate these skills with others you already have so that you will become a more effective interviewer.

The following client statements are derived from the interview you've been following. Develop the response specified for each of these statements and write it on a separate worksheet.

1. Perhaps when I consider his behavior, I regret adopting him, but given the enjoyment we've had over the years, I'm glad we did it. (paraphrase)
2. Yes. If we'd just had some indication that it was happening. But all of a sudden, the police officer was at the door. And it's not just one incident, but 14. (reflection of feeling)
3. I guess if I look at it, the problem has been developing over the past 3 or 4 years. He started showing up late for school. . . . (minimal encouragement)
4. I don't know. He stopped doing what we asked him to do. (open question)
5. Over the past couple of summers, he and I have agreed that he'd mow the lawn and get paid for it. Well, he's done it for a couple of weeks each summer, and then he just stops. It's things like that. What I consider to be his responsibilities around the house aren't responsibilities as far as he's

concerned. He thinks he's just a boarder. (summarization, covering responses 1 through 5)

The following are examples of possible responses to the preceding statements:

1. a. The pleasure you've experienced over the years far outweighs the problems you're having now.
   b. Generally, you have few regrets about having adopted him, although your present difficulties have caused you to question that decision.
2. a. Finding out that way overwhelmed you.
   b. The suddenness and enormity of the problem seemed to alarm you.
3. a. Hmmm . . . hmmm. . . .
   b. And . . .
4. a. Can you give me an example of that?
   b. What other things caused you concern?
5. a. So, to summarize what you've told me up to this point, your son has given you a lot of joy, and although you'd been concerned about his behavior, you were unprepared for this visit from the police.
   b. Let me see if I understand you correctly. Although you'd been concerned about your son's behavior, you were unprepared for this visit from the police. You've had a good relationship with him until recently.

Points to Remember about Integrating Skills

1. WHEN ATTENDING, USING OPEN QUESTIONS, GIVING MINIMAL ENCOURAGEMENT, AND REFLECTING FEELING AND CONTENT:
   a. Master each skill to the best of your ability. Effective integration of interviewing skills depends on your ability to use each individual skill.
2. WHEN USING SINGLE SKILLS IN AN INTERVIEW:
   a. Respond to clients in a way that is comfortable for you.
   b. Combine single skills when necessary.
   c. Don't overuse a particular skill or a limited range of skills.
3. THE SKILLS COVERED THUS FAR ARE APPROPRIATE:
   a. In response to most client statements, as long as they demonstrate good attending.

6.1

Activity Unit

Find two others who have also read this chapter. Role play short interviews (10 to 15 interviewer responses) alternating the client, interviewer, and observer roles. Record each brief interview on audio- or videotape.

Interviewer role Practice maintaining appropriate eye contact, vocal behavior, professionally appropriate posture, and listening throughout the interview. Also practice using open questions, minimal encouragement, reflection of feeling, paraphrases, or summarization as they seem appropriate in response to the client. When it is important to obtain specific information, use closed questions. Combination responses are also appropriate, such as a paraphrase and an open question.

Client role Be cooperative and provide sufficient information for the interviewer to practice. Clients may want to role play the problem of a friend or family member or to enact one of the following roles:

1. An individual visiting his or her family for the first time in 3 or 4 years.
2. A widow or widower grieving the loss of her or his spouse.
3. An employee who fears she is developing a drinking problem.
4. A person discussing reactions to an upcoming change in residence from the East to the West Coast.

Observer role Operate the equipment and complete the Practice Interview Checklist (see Table 6.1).

Table 6.1

Practice Interview Checklist

Activity	Interviewer response[a]														
	1	2	3	4	5	6	7	8	9	10	11	12	13	14	15
Effective nonverbal behavior															
On topic															
Open question															
Closed question															
Minimal encouragement															
Reflection of feeling															
Paraphrase															
Summarization															
Effective response[b]															

[a] Check the activities that characterize each response.
[b] Indicate the degree of effectiveness on a scale from 1 to 5 in which 1 is ineffective and 5 is very effective.

Postinterview discussion Using the taped interview and the Practice Interview Checklist data, the participants should discuss the performance of the interviewer. All participants can learn from this discussion, which should be nonjudgmental, focus on positive as well as less effective responses, and stimulate improved performance for all.

6.2

Activity Unit

Using the tape of the interview produced in Activity Unit 6.1 or the tape of another practice interview, enter the original interview responses below. Then formulate a better response to replace each original one in the space provided.

Original response 1 _____

_____

_____

Better response _____

_____

_____

Original response 2 _____

_____

_____

Better response _____

_____

_____

Original response 3 _____

_____

_____

Better response _____

_____

_____

Original response 4 _____

_____

_____

Better response _____

_____

_____

Original response 5 _____

_____

_____

Better response _____

_____

_____

Original response 6 _____

_____

_____

Better response _____

_____

_____

Original response 7 _____

_____

_____

Better response _____

_____

_____

Original response 8 _____

_____

_____

Better response _____

_____

_____

Original response 9 _____

_____

_____

Better response _____

_____

_____

Original response 10 _____

_____

_____

Better response _____

_____

_____

Original response 11 _____

_____

_____

Better response _____

_____

_____

Original response 12 _____

_____

_____

Better response _____

_____

_____

Original response 13 _____

_____

_____

Better response _____

_____

_____

Original response 14 _____

_____

_____

Better response _____

_____

_____

Original response 15 _____

_____

_____

Better response _____

_____

_____

Cultural Considerations

1. Remember that the issue the client brings to the session is developed in a cultural context, and the goal is to help the client resolve the issue within that context. It is important that the interviewer listen for family and cultural issues that affect the client.
2. When interviewing clients from other cultures, the interviewer should be flexible and adapt his or her use of skills to accommodate the client's culture.

InfoTrac® College Edition Keyword Search Terms

Skills AND Counseling

Chapter Seven

Communicating Feeling and Immediacy

$\mathcal{I}$n Chapters 1 through 6, we focused on basic interviewing skills that enable you to develop a good working relationship and open communication with a client. In Chapters 7 through 13, we focus on advanced interviewing skills—communicating feeling and immediacy, confronting, self-disclosing, information giving, interpreting, structuring, and enlisting cooperation—that enable you to help a client focus on problems and act on them. Chapter 14 helps you integrate the basic and advanced skills into an effective interviewing style.

This chapter is intended to help you master the skills of communicating feeling and immediacy. After completing this chapter, you should be able to:

1. Identify your feelings as you interact with a client.
2. Communicate your feelings to a client.
3. Decide whether communication of feelings is appropriate in the situation.
4. Focus the communication between yourself and a client on immediate concerns.
5. Decide when it's appropriate to focus an interview on immediate concerns.

Communicating feeling is an important aspect of immediacy. It can involve interviewers sharing their feelings about the relationship with the client, or their reaction to the client's behavior, what the client has said or what the client has left unsaid. Immediacy involves attending to issues in the interview that, if ignored, could interfere with the relationship or the progress of the interview. Communicating feeling and immediacy is primarily used to build trusting, open relationships and to facilitate the resolution of potentially disruptive aspects of the interview. The use of immediacy in the interview can provide a learning experience that clients can extend to their own interpersonal relationships.

The ability to recognize your own feelings increases your ability to identify the feelings of others. The first section of this chapter helps you identify your own feelings. Then you will be presented with material that helps you communicate your feelings and concerns about what is occurring in the immediate relationship between you and the client.

Most of the frames in this chapter are complete in themselves; that is, each frame focuses on a different client. Go to 7.1.

| **7.1** | *People often find it difficult to identify and communicate their feelings. During an interview, you should be able to identify your feelings and communicate them to a client. Communicating feeling is similar to reflecting feeling, except that the emphasis is on the interviewer's feelings rather than on the client's.* |

CLIENT: I was just getting into my job, and then they let me go. When I first applied for the job, I knew it was for me. When they called to tell me I had it, I was so excited. I really worked hard at it, and everyone said I was doing so well. . . . I really tried. Last Friday, the manager called me in, gave me my severance pay, and said he was sorry, but cutbacks made it necessary for him to let me go.

INTERVIEWER: I'm glad you were able to land the job. That's what's important. (Go to 7.2)

INTERVIEWER: Damn! Some companies make me so mad. (Go to 7.3)

INTERVIEWER: I'm really sorry to hear about your misfortune. (Go to 7.4)

7.2   Your answer: *I'm glad you were able to land the job. That's what's important.*
You may be glad to hear of the client's success in obtaining a job; however, the expression of this emotion is inappropriate because the client no longer has the job. This response suggests that you aren't listening to the client's entire message. Return to 7.1, attend to the client's message and your own feelings, and then respond with an appropriate expression of feeling.

7.3   Your answer: *Damn! Some companies make me so mad.*
This response is excessive and far too general. When you express your feelings, the intensity of your statement should be appropriate to the particular situation. Moreover, the response you've selected changes the topic from the

client to a general discussion of the unfairness of some companies and their personnel policies. Return to 7.1 and try again.

7.4   **Your answer:**   *I'm really sorry to hear about your misfortune.*
Correct. Given the nature of the client's disappointing experience, this is an appropriate expression of feeling. Go to 7.5.

7.5

An important aspect of communicating feelings is the ability to identify and monitor them. As you listen attentively to a client, be aware of your reactions to what is being said. A direct but moderated expression of feeling can help a client effect change. No matter what your reaction, your response must always be respectful of the client and the interviewing relationship.

CLIENT: (having failed to bring important information concerning his problem) I'm sorry. I left my records at home again. They aren't that important, are they? I'm sure I'll remember next week.

Choose the most appropriate response.

INTERVIEWER: I'm upset that you forgot your records. It's the only way we can evaluate your progress. (Go to 7.6)

INTERVIEWER: Not again. Well, I guess we can work around it. (Go to 7.7)

INTERVIEWER: How are we supposed to get anywhere? Clients who don't cooperate make me angry. (Go to 7.8)

7.6   **Your answer:**   *I'm upset that you forgot your records. It's the only way we can evaluate your progress.*
Correct. You've communicated your feelings in a way that will be useful to the client. Always remember to express your feelings in a manner that respects the client and the relationship. Go to 7.9.

7.7   **Your answer:**   *Not again. Well, I guess we can work around it.*
The initial part of your response suggests you aren't sharing all of your feelings with the client. To effect some change in the client's behavior, it is necessary that you share these feelings. Otherwise, they will interfere with your efforts to establish an open relationship with the client. Return to 7.5 and try again.

7.8   **Your answer:**   *How are we supposed to get anywhere? Clients who don't cooperate make me angry.*
You've expressed not only your anger toward this client but also the anger you've accumulated from your interactions with other clients. Note your reactions to

each client and express your feelings in a careful and facilitative manner that respects the client and the relationship. Return to 7.5 and try again.

| 7.9 | | *Be aware of your bodily reactions (clenched jaw, frown, finger or foot tapping, and the like) as well as your thoughts and feelings.* |

CLIENT: (a student who has come to discuss poor grades) Last Saturday, the apartment manager found me cooking hot dogs and told me they smelled terrible. How can she think that what I cook is causing the smells? When did hot dogs ever cause smells? I don't think I'm the culprit.

Choose the most appropriate response.

INTERVIEWER: (fidgeting and looking out the window while the client is talking) It seems you're having real difficulties with the apartment manager. (Go to 7.10)

INTERVIEWER: (fidgeting and looking out the window while the client is talking) I get angry when people smell up a place with food. (Go to 7.11)

INTERVIEWER: (fidgeting and looking out the window while the client is talking) I feel quite distracted by what you're saying, and I'm puzzled about how it relates to your problem. (Go to 7.12)

7.10   Your answer: *(fidgeting and looking out the window while the client is talking) It seems you're having real difficulties with the apartment manager.*
The client has come to talk about poor grades. Because it is very difficult for the client to discuss that problem, the client may focus on an irrelevant issue. You've responded to the client's inappropriate discussion. Return to 7.9 and try again.

7.11   Your answer: *(fidgeting and looking out the window while the client is talking) I get angry when people smell up a place with food.*
You've ignored your fidgeting and expressed an irrelevant bias that will probably lead to further discussion of hot dogs and cooking. Recognize your bodily reactions and share your feelings with the client so that they can be utilized. Return to 7.9 and try again.

7.12   Your answer: *(fidgeting and looking out the window while the client is talking) I feel quite distracted by what you're saying, and I'm puzzled about how it relates to your problem.*
Correct. This response effectively communicates your feelings to the client. Go to 7.13.

| 7.13 | | *In Chapter 2, we stressed the importance of maintaining appropriate eye contact and a relaxed, professional posture while listening to a client. Eye contact and posture are also important when you communicate information to a client.* |

CLIENT: (discussing improved behavior of her son) I'm really glad I came to see you. John was just unmanageable when I first came, but now he's absolutely no problem. He's doing well at school. I don't know how I can thank you. . . . It's all due to you.

Choose the most appropriate response.

INTERVIEWER: (relaxed and looking warmly at the client) It makes me feel good to hear you say that. (Go to 7.14)

INTERVIEWER: (fidgeting and looking away from the client) It makes me feel . . . really good to hear you say that. (Go to 7.15)

INTERVIEWER: (looking directly at the client) You did all the work. (Go to 7.16)

7.14 Your answer: *(relaxed and looking warmly at the client) It makes me feel good to hear you say that.*
Correct. Your posture, eye contact, and mode of delivery complement your verbal message and indicate to the client that you are happy to hear her remarks. Go to 7.17.

7.15 Your answer: *(fidgeting and looking away from the client) It makes me feel . . . really good to hear you say that.*
This reaction may confuse and disconcert the client. Use your body and your eyes to complement your verbal message while speaking directly to the client. Return to 7.13 and try again.

7.16 Your answer: *(looking directly at the client) You did all the work.*
When interviewers are embarrassed by a client's positive or complimentary remarks, they sometimes contradict the client. When this happens, the client is denied the right to express positive feelings. As an interviewer, be aware of your reactions to positive feedback. Return to 7.13 and try again.

7.17   *While maintaining appropriate eye contact and a relaxed posture, keep your communication of feelings on topic.*

CLIENT: (discussing failure with a weight-reducing program) I really have been doing all the things we decided on, but I'm just not losing any weight. And I'm so afraid that if I don't lose weight soon, I'll have another heart attack. I'm getting so depressed. I'm doing all the things I'm supposed to do, but it just isn't working. . . . sighing) I don't know.

Choose the most appropriate response.

INTERVIEWER: I'm worried about a number of things that are going on here. I guess we just have to go on trying. (Go to 7.18)

INTERVIEWER: What you're saying concerns me very much. (Go to 7.19)

INTERVIEWER: I'm pleased to see you today. I wasn't sure you would come. (Go to 7.20)

7.18 **Your answer:** *I'm worried about a number of things that are going on here. I guess we just have to go on trying.*
Your response should show that you are listening to the client and are interested in what he or she has to say. This response ignores the client's concerns and offers false reassurance. Return to 7.17 and try again.

7.19 **Your answer:** *What you're saying concerns me very much.*
Correct. You've demonstrated that you understand the importance of communicating feelings that are on topic. Go to 7.21.

7.20 **Your answer:** *I'm pleased to see you today. I wasn't sure you would come.*
This response is off topic. Focus on what the client is saying. Return to 7.17 and try again.

7.21

When interviewers communicate their feelings, they clarify their reactions, demonstrate how to recognize and express feelings, and help develop trusting, open relationships with their clients.

CLIENT: (a widow who has been seeing her minister regularly to discuss her husband's recent death) I got it! I got it! I just had to come and tell you. I actually got the job. I start on Monday.

Choose the most appropriate response.

INTERVIEWER: I'm so surprised. I wasn't sure that you'd be able to do it. (Go to 7.22)

INTERVIEWER: I'm so relieved. It was beginning to bother me that you seemed to be getting nowhere. (Go to 7.23)

INTERVIEWER: I'm so glad . . . so happy for you. I think that's the greatest news. (Go to 7.24)

7.22 **Your answer:** *I'm so surprised. I wasn't sure that you'd be able to do it.*
Your response is a communication of feeling, but not one that is likely to build a trusting, open relationship. The client is likely to view your surprise as a put-down. Return to 7.21 and try again.

7.23 **Your answer:** *I'm so relieved. It was beginning to bother me that you seemed to be getting nowhere.*
What the client has just told you is likely to produce feelings of happiness, which are the most appropriate to express at the moment. The communication of your negative feelings won't serve to develop a trusting, open relationship. Return to 7.21 and try again.

7.24 **Your answer:** *I'm so glad . . . so happy for you. I think that's the greatest news.*
Correct. The client's eagerness to share this news with you suggests a good relationship, and your communication of positive feelings will further this relationship. Proceed to 7.25.

7.25	

When you, as an interviewer, suspect that your biases and feelings are interfering with your relationship with the client, inform the client of this and make alternative arrangements if the client wishes.

CLIENT: (discussing her feelings about her pregnancy) Our first child has mental retardation. The doctor thinks that it's genetic and that I should have an abortion. That's easy for him to say. . . . I'm just not comfortable with the idea. I can't get it out of my mind. I know it would be hard to raise two children with mental retardation. Maybe it would be better for us all if I were to have an abortion. . . . I just don't know.

Choose the most appropriate response.

INTERVIEWER: Abortion may be the only answer to your problem. (Go to 7.26)

INTERVIEWER: Hearing of your difficult decision makes me feel really concerned for you. (Go to 7.27)

INTERVIEWER: I'm really concerned to hear you talk about abortion. I get upset when people encourage abortions. (Go to 7.28)

7.26 Your answer: *Abortion may be the only answer to your problem.*

This statement suggests that you have strong feelings in favor of abortion and have permitted them to influence your response. Return to 7.25 and try again.

7.27 Your answer: *Hearing of your difficult decision makes me feel really concerned for you.*

Correct. This response communicates unbiased feelings concerning the client's problem. Everyone possesses bias. It is important that interviewers recognize their own bias and act to ensure that the client is not influenced by them. Go to 7.29.

7.28 Your answer: *I'm really concerned to hear you talk about abortion. I get upset when people encourage abortions.*

This response communicates your personal bias. If you do feel antagonistic toward the idea of abortion, your views will interfere with your interaction with the client. Encourage the client to seek counseling elsewhere. Return to 7.25 and select a more appropriate response.

7.29	

Immediacy—the ability to discuss directly and openly what is happening in a relationship—is closely related to the skill of communicating feeling.

CLIENT: (a young woman who wants to leave home is having difficulty achieving this goal) I really don't know where to begin today. (pauses and glances out the window) I chatted with my mother about finding a place of my own. (glances out the window) We had a long talk. (looks down at the floor) It was a good talk. (fidgets and looks down at the floor)

Choose the most appropriate response.

INTERVIEWER: Do you think you'll be moving soon, then? (Go to 7.30)

INTERVIEWER: I get the feeling that you're having difficulty telling me what occurred. (Go to 7.31)

INTERVIEWER: It sounds as though you had quite a good talk. (Go to 7.32)

7.30   Your answer: *Do you think you'll be moving soon, then?*
The client's behavior suggests that she is having some difficulty talking to you. You should resolve this difficulty to make her feel more comfortable. Return to 7.29 and try again.

7.31   Your answer: *I get the feeling that you're having difficulty telling me what occurred.*
Correct. By focusing on the discomfort the client is experiencing, you've encouraged her to discuss this discomfort and explore its meaning. Go to 7.33.

7.32   Your answer: *It sounds as though you had quite a good talk.*
You've accurately reflected the content of the client's statement, but you haven't focused on her behavior and your reaction to it. By focusing on the meaning of her behavior, you can help the client explore her problem more directly. Return to 7.29 and try again.

7.33   *To engage in immediacy, you must focus not only on what is being said but also on what is left unsaid. For example, you might focus on the way a client expresses himself or herself, the feelings that develop between you and the client, or the nonverbal behavior that accompanies your discussion. When you engage in immediacy, respond in the present tense.*

CLIENT: (referred to a psychologist because of difficulty with social relationships) Gee, am I ever mad. I went to the employment office on Tuesday—nothing. And I didn't get anywhere by looking in the paper. I've been coming here for 3 months, and I still don't have a job. What the hell are you doing? I came here so that you could help me get a job.

Choose the most appropriate response.

INTERVIEWER: From what you've said, it seems that you made some effort to find a job but were disappointed again. (Go to 7.34)

INTERVIEWER: From what you're saying, I get the feeling that you don't think I'm doing my job. It seems that I'm not doing what you want me to do. (Go to 7.35)

INTERVIEWER: As I remember, you weren't very comfortable with our decision of last week. (Go to 7.36)

7.34   Your answer: *From what you've said, it seems that you made some effort to find a job but were disappointed again.*
You've attended to the content of the client's statement, but you've ignored an area of greater concern—the immediate feelings between you and the client.

Until you help the client communicate and resolve his hostility toward you, there can be little progress toward resolving his problems. Return to 7.33 and try again.

7.35	**Your answer:**	*From what you're saying, I get the feeling that you don't think I'm doing my job. It seems that I'm not doing what you want me to do.*

Correct. This is a good response in which you communicate, in the present tense, the message underlying the client's statement. Go to 7.37.

7.36	**Your answer:**	*As I remember, you weren't very comfortable with our decision of last week.*

One important aspect of immediacy is dealing with what is going on at the present moment in a relationship. This response deals with issues that were important in the past. Return to 7.33 and try again.

7.37

The client is likely to engage in frank, open discussion when you use immediacy. Therefore, you must be willing to converse in an open, nondefensive manner.

CLIENT: (continuing from the last interaction) You're damned right. As the weeks go by, I'm getting more and more scared that I'll never get a job again. And all you do is give me these silly assignments. You aren't helping me a bit.

Choose the most appropriate response.

INTERVIEWER: I don't think we should discuss what I do. Let's get back to your problem. (Go to 7.38)

INTERVIEWER: You just don't do what I ask you to do. (Go to 7.39)

INTERVIEWER: You're becoming so anxious that you're afraid I won't be able to help you after all. (Go to 7.40)

7.38	**Your answer:**	*I don't think we should discuss what I do. Let's get back to your problem.*

When you focus on the immediate situation, you must be prepared to help the client in a frank and open discussion. By being defensive, you haven't allowed the client to deal with the hostility that is preventing him from cooperating with you. Return to 7.37 and try again.

7.39	**Your answer:**	*You just don't do what I ask you to do.*

When you counterattack, the client will either back off or come back with an even more aggressive response. In either case, the client will assume that it's impossible to carry on a frank and open discussion of his immediate feelings with you. Remember, when you initiate immediacy, you must be ready to follow through with open discussion. Return to 7.37 and try again.

7.40	**Your answer:**	*You're becoming so anxious that you're afraid I won't be able to help you after all.*

Correct. By focusing in an open and nondefensive manner on what the client has said, you've encouraged further discussion. Go to 7.41.

7.41

Discussing what is occurring in a relationship is often a new and difficult experience for the client. If you are too direct, the client may withdraw from the discussion, but if you are careful and sensitive, the client will probably feel comfortable with you.

CLIENT: (a woman who has established a positive relationship with the therapist) I just can't get to sleep at night. . . . I keep having these awful thoughts. I know you're going to ask me what they are, but I can't tell you. I get along well with you, but if I told you, you might not want to see me anymore. . . . Can't you just tell me what to do?

Choose the most appropriate response.

INTERVIEWER: You feel that we share a good relationship, but that discussing these thoughts with me could alter it. (Go to 7.42)

INTERVIEWER: You feel that we share a good relationship—how good is it if you can't trust me? (Go to 7.43)

INTERVIEWER: If you feel that we share a good relationship, why can't you tell me about these thoughts? (Go to 7.44)

7.42   Your answer: *You feel that we share a good relationship, but that discussing these thoughts with me could alter it.*
Correct. This response helps the client feel more comfortable discussing a sensitive subject. Go to 7.45.

7.43   Your answer: *You feel that we share a good relationship—how good is it if you can't trust me?*
Because discussing this subject with you is obviously difficult for the client, your response is far too direct. To help the client, be cautious in your response. Return to 7.41 and try again.

7.44   Your answer: *If you feel that we share a good relationship, why can't you tell me about these thoughts?*
The client has told you that she's finding it difficult to discuss these thoughts with you. This attempt to shame her into revealing her thoughts will probably cause her to feel guilty and defensive. Your response should invite the client to discuss the reasons behind her reluctance to reveal her thoughts. Return to 7.41 and try again.

7.45

When the interviewer engages in immediacy, the client develops a sensitivity to his or her own behavior and its effect on other people.

CLIENT: (having previously made similar comments) From what I said before, I'm sure you have a bad impression of my wife. I really didn't mean to portray her like that. She's actually a very good person. I hope you don't think badly of her.

Choose the most appropriate response.

INTERVIEWER: Of course not. I'm sure she's a very good person. (Go to 7.46)

	INTERVIEWER:	You seem to be content with your wife. (Go to 7.47)
	INTERVIEWER:	You often find it necessary to apologize for your comments. It seems as though you have difficulty accepting responsibility for your own statements. (Go to 7.48)

7.46 Your answer: *Of course not. I'm sure she's a very good person.*

By focusing on the content of the client's message, you reinforce his behavior. It is important that the client take responsibility for his own comments and feelings. By focusing on his present behavior and talking about it, you will help him realize what he is doing. Return to 7.45 and try again.

7.47 Your answer: *You seem to be content with your wife.*

Although this response may placate the client, it won't help him develop the ability to communicate directly and honestly. Return to 7.45 and choose a response that focuses on your observations of the interaction between you and the client.

7.48 Your answer: *You often find it necessary to apologize for your comments. It seems as though you have difficulty accepting responsibility for your own statements.*

Correct. This response gives the client feedback on how he expresses himself and invites him to reconsider his responsibility for his manner of expression. Go to 7.49.

7.49

When the client engages in immediacy, discuss his or her concerns directly.

CLIENT: I keep coming here and discussing the same things week after week. I'm sure you must be getting fed up with me.

Choose the most appropriate response.

INTERVIEWER: Of course I'm not getting fed up with you. Could you tell me what's been happening to you this week? (Go to 7.50)

INTERVIEWER: You're not sure you're getting anywhere, and you're wondering what I think about that. (Go to 7.51)

INTERVIEWER: That's really not important. I'm here to help you. (Go to 7.52)

7.50 Your answer: *Of course I'm not getting fed up with you. Could you tell me what's been happening to you this week?*

When the client expresses a concern about the immediate situation, you need to enter into a discussion of this issue. Your response should demonstrate your willingness to discuss the client's immediate concerns about his or her progress. Return to 7.49 and try again.

7.51 Your answer: *You're not sure you're getting anywhere, and you're wondering what I think about that.*

Correct. This response focuses directly on the client's immediate concern, encouraging discussion of this issue. Go to 7.53.

7.52 **Your answer:** *That's really not important. I'm here to help you.*
The client has raised a concern that requires your immediate attention. Your response not only ignores the client's concern but also belittles it. When a client focuses on the interaction between himself or herself and the interviewer, it is the responsibility of the interviewer to assist in the expression of this concern. Return to 7.49 and try again.

7.53 *Differences in age, gender, race, viewpoints, religion, and socioeconomic background often affect the degree of compatibility between the interviewer and the client and give rise to situations in which the interviewer needs to use immediacy responses.*

CLIENT: (an older man who finds himself in the presence of a young interviewer) I don't feel quite right discussing my problem with you. How long have you been doing this work?

Choose the most appropriate response.

INTERVIEWER: Because I'm young, you're wondering whether I have enough experience to be able to help you. (Go to 7.54)

INTERVIEWER: I'm sure I can help you. Could you tell me what brings you here? (Go to 7.55)

INTERVIEWER: My age has nothing to do with my abilities. I have excellent credentials. (Go to 7.56)

7.54 **Your answer:** *Because I'm young, you're wondering whether I have enough experience to be able to help you.*
Correct. You've identified the client's immediate concern and encouraged frank and open discussion of his worry about your age. Go to 7.57.

7.55 **Your answer:** *I'm sure I can help you. Could you tell me what brings you here?*
You are likely to gain the confidence of the client only after you have responded to his immediate concern and have helped to resolve it. Return to 7.53 and try again.

7.56 **Your answer:** *My age has nothing to do with my abilities. I have excellent credentials.*
Although you've identified the client's concern, your authoritative and defensive tone will probably alienate him further. Your response should invite the client to discuss his concern in a frank and open manner. Return to 7.53 and try again.

7.57 *Immediacy responses can establish trust between the interviewer and the client.*

CLIENT: I don't know whether I want to answer that question. After all, you're the personnel manager, and this information is highly personal. How do I know what you'll do with it?

Choose the most appropriate response.

INTERVIEWER: I can assure you that everything you say here will be held in confidence. (Go to 7.58)

INTERVIEWER: I didn't realize that my question would have such an effect on you. You should have nothing to hide from me. (Go to 7.59)

INTERVIEWER: You find it difficult to trust me with this information. (Go to 7.60)

7.58   Your answer: *I can assure you that everything you say here will be held in confidence.*
Whenever there is a question concerning the degree to which either the interviewer or the client can be trusted, it is important that the issue of trust be considered immediately. Return to 7.57 and try again.

7.59   Your answer: *I didn't realize that my question would have such an effect on you. You should have nothing to hide from me.*
You've responded to the immediate situation between you and the client, but your response implies that the client does indeed have something to hide. To gain the client's trust, focus on the issue of trust in a more tentative, nonthreatening manner. Return to 7.57 and try again.

7.60   Your answer: *You find it difficult to trust me with this information.*
Correct. This response encourages the client to discuss the issue of trust. Go to 7.61.

7.61   Some clients become overdependent; others resist the interviewer's help. Immediacy responses are appropriate when the degree of dependency becomes a cause of concern for the client or the interviewer.

CLIENT: I did all the things you suggested last week, and they worked well. But some other things have just come up, and I need you to tell me what to do about them.

Choose the most appropriate response.

INTERVIEWER: Why don't you tell me about them so we can decide which one is most important. (Go to 7.62)

INTERVIEWER: I wonder if you could apply some of the things you learned last week to these problems and solve them yourself. (Go to 7.63)

INTERVIEWER: I'm concerned about the fact that you see me as the one who solves your problems. (Go to 7.64)

7.62   Your answer: *Why don't you tell me about them so we can decide which one is most important.*
This response suggests that you are comfortable when the client is dependent on you. To establish an equal working relationship with the client, you must engage in an open, frank discussion of the dependency issue. Return to 7.61 and try again.

7.63   Your answer: *I wonder if you could apply some of the things you learned last week to these problems and solve them yourself.*

This is a laudable goal, but without some prior discussion of the dependency issue, your response will probably frighten the client. Return to 7.61 and try again.

7.64   **Your answer:**   *I'm concerned about the fact that you see me as the one who solves your problems.*
Correct. This response focuses on the fact that the client depends on you to solve problems. Moreover, it invites frank and open discussion of this issue. Go to 7.65.

7.65   *Immediacy responses are called for when the client engages in aimless, circular discussion or agrees to achieve a particular goal but does nothing to attain it.*

CLIENT: (having spent the first 10 minutes of the session discussing his uneventful week) It's been a pretty quiet week. I guess I just sat around most of the time.

Choose the most appropriate response.

INTERVIEWER: As I listen to you, I find myself wondering why we're not discussing more important things. (Go to 7.66)

INTERVIEWER: What else would you like to talk about today? (Go to 7.67)

INTERVIEWER: I don't think this is getting us anywhere. Why don't you come back in another week? (Go to 7.68)

7.66   **Your answer:**   *As I listen to you, I find myself wondering why we're not discussing more important things.*
Correct. By raising your concern about the apparent lack of direction in the client's statements, you've opened the way for an open, relevant discussion. Go to 7.69.

7.67   **Your answer:**   *What else would you like to talk about today?*
This response invites more aimless chatter. Your response should focus on your concern and invite an open discussion of it. Return to 7.65 and try again.

7.68   **Your answer:**   *I don't think this is getting us anywhere. Why don't you come back in another week?*
Although this response focuses on your immediate concern, it's too abrupt to be of any benefit. Return to 7.65 and try again.

7.69   *Clients frequently experience some discomfort during the initial and final stages of a relationship with an interviewer. When this happens, the interviewer should focus on these feelings and help the client resolve them.*

CLIENT: I won't be coming here anymore. . . . Won't that be strange? I'm used to coming here. . . . It's been so good. But I guess you'll be glad to get rid of me.

Choose the most appropriate response.

INTERVIEWER: You've been seeing me for a long time, but you're well in control now. (Go to 7.70)

INTERVIEWER: We've been meeting for a long time, and we have a good relationship. I guess

	you're telling me that you'll miss me. I'll miss you too. (Go to 7.71)
INTERVIEWER:	I won't be glad to get rid of you, but you don't need me anymore. You're strong enough now to manage things by yourself. (Go to 7.72)

7.70	**Your answer:**	*You've been seeing me for a long time, but you're well in control now.*
		This response is somewhat cold and abrupt. Your response should encourage the discussion of feelings about the conclusion of your relationship with the client. Return to 7.69 and try again.

7.71	**Your answer:**	*We've been meeting for a long time, and we have a good relationship. I guess you're telling me that you'll miss me. I'll miss you too.*
		Correct. This response, combining immediacy and expression of feeling, encourages the client to engage in discussion of immediate concerns. Go to 7.73.

7.72	**Your answer:**	*I won't be glad to get rid of you, but you don't need me anymore. You're strong enough now to manage things by yourself.*
		When the relationship between the interviewer and the client has been a good one, separation is often difficult for the client. Your response prohibits any discussion of the client's feelings, as well as any possibility of resolving them. Return to 7.69 and try again.

7.73

Review Questions

Indicate whether each of the following statements is true or false.

1. The interviewer's communication of feeling facilitates the client's communication of feeling.
2. It's appropriate for the interviewer to communicate all his or her feelings to the client.
3. To use immediacy responses appropriately, the interviewer needs to attend to the client's verbal message.
4. To communicate feeling, the interviewer must be aware not only of the client's message but also of his or her own reaction to that message.
5. Immediacy focuses on what is happening in the here and now of a relationship.
6. Direct, rather than tentative, communications of immediacy lead to frank and open discussion.
7. The interviewer should be the only one to initiate discussion of the relationship.
8. Discussion of the issue of trust between the interviewer and the client can be damaging to their relationship.
9. Interpersonal relationships are enhanced by the skill of immediacy.
10. When the interviewer initiates immediacy, he or she should be prepared to enter into a direct, honest discussion.

Review Answers
1. *True.* If you answered *false*, go to 7.21 and review.
2. *False.* If you answered *true*, go to 7.25 and review.
3. *False.* If you answered *true*, go to 7.33 and review.
4. *True.* If you answered *false*, go to 7.5 and review.
5. *True.* If you answered *false*, go to 7.29 and review.
6. *False.* If you answered *true*, go to 7.41 and review.
7. *False.* If you answered *true*, go to 7.49 and review.
8. *False.* If you answered *true*, go to 7.57 and review.
9. *True.* If you answered *false*, go to 7.45 and review.
10. *True.* If you answered *false*, go to 7.37 and review.

If seven or more of your responses were incorrect, you should return to 7.1 and work through this chapter again.

Points to Remember about Communicating Feeling and Immediacy

1. WHEN IDENTIFYING YOUR FEELINGS:
    a. Identify the feelings evoked by your thoughts about what the client is communicating.
    b. Identify the feelings associated with your bodily reactions to the client's message.
2. WHEN COMMUNICATING YOUR FEELINGS:
    a. Maintain appropriate eye contact and a relaxed, professional posture.
    b. Communicate feelings that are relevant to the topic being discussed.
    c. Use an appropriately moderated form of expression.
3. COMMUNICATE FEELINGS:
    a. When you want to clarify your reactions for the client, act as a model for the client, or develop a trusting, open relationship with the client.
    b. When your feelings are provoked by an unbiased response to the client's message.
4. WHEN COMMUNICATING IMMEDIATE CONCERNS:
    a. Focus both on what is being said and also on what is not being said.
    b. Respond in the present tense.
    c. Express yourself carefully and sensitively.
    d. Be ready to follow up openly and nondefensively.
5. COMMUNICATE IMMEDIATE CONCERNS TO:
    a. Promote direct mutual communication.
    b. Resolve tensions and discomforts.
    c. Focus on and resolve incompatibilities.
    d. Clarify the issues concerning trust.
    e. Focus on and resolve the client's dependency.
    f. Resolve circular discussion and client inactivity.

g. Resolve the client's feelings during the initial and final stages of the interviewing process.

7.1 Activity Unit

Find two others who have also read this chapter. Role play brief interviews alternating the client, interviewer, and observer roles. Record each brief interview on audio- or videotape. An interview of four to five interviewer responses is sufficient for this exercise.

Interviewer role Practice communicating feeling and immediacy in response to the client.

Client role Be cooperative and provide sufficient information for the interviewer to practice. Clients may want to role play the problem of a friend or family member or to enact one of the following roles:

1. A client, concerned about a mismatch in age, gender, or race, who becomes angry with the interviewer and indicates that the interviewer does not understand her or him.
2. An overweight client, who has not kept records for several weeks, is angry with the therapist for his or her lack of progress, and is wondering whether to terminate contact.
3. A client who has just obtained a promotion that requires a change in residence and fears losing contact with the interviewer.
4. An angry, hostile client who has been ordered by the court to see the interviewer. The client believes no problem exists, and that he or she should not have to be interviewed.

Observer role Operate the equipment and complete the Practice Interview Checklist (see Table 7.1).

Postinterview discussion Using the taped interview and the Practice Interview Checklist data, the participants should discuss the performance of the interviewer. All participants can learn from this discussion, which should be nonjudgmental, focus on positive as well as less effective responses, and stimulate improved performance for all.

Table 7.1

Practice Interview Checklist

Activity	Interviewer response[a]				
	1	2	3	4	5
Appropriate nonverbal behavior					
On topic					
Communication of feeling					
Communication of immediacy					
Effective response[b]					

[a] Check the activities that characterize each response.
[b] Indicate the degree of effectiveness on a scale from 1 to 5 in which 1 is ineffective and 5 is very effective.

7.2

Activity Unit

Using the tape of the interview produced in Activity Unit 7.1 or the tape of another practice interview, enter the original interview responses below. Then formulate a better response to replace each original one in the space provided.

Original response 1 _____

_____

Better response _____

_____

Original response 2 _____

_____

Better response _____

_____

Original response 3 _____

_____

Better response _____

_____

_____

Original response 4 _____

_____

Better response _____

_____

_____

Original response 5 _____

_____

_____

Better response _____

_____

_____

Cultural Considerations

1. For clients of some cultures, it is important to communicate feeling and focus on immediate concerns in a manner that permits the client to save face.
2. When clients of other cultures perceive the interviewer as possessing cultural biases or when cultural differences become an issue, it is important that the skill of immediacy be implemented to resolve such biases and differences before the interview continues.

Chapter Eight

Confronting

*T*his chapter is intended to help you master the skill of confronting. After completing this chapter, you should be able to:

1. Identify discrepancies in a client's communication.
2. Confront a client with discrepancies you've identified.
3. Explain the purpose of confrontation.
4. Decide whether confronting is appropriate in the situation.

In common usage, *confrontation* is often seen as hostile or punitive. In interviewing, confrontation is a high-level skill in which the interviewer draws to the attention of the client discrepant aspects of the client's verbal or non-verbal behavior. Confrontation is used to assist clients to identify and consider discrepancies in their thoughts, behaviors, and feelings in order to resolve their problems and issues. When used effectively, confrontation helps clients to make changes in their lives.

Confrontation should be tentative and nonjudgmental, encouraging the client to explore the discrepancy. Once a confrontation is made, the interviewer should use previously learned skills to facilitate discussion of and resolution of discrepancies. Like communications of feeling and immediacy, confrontation increases a client's awareness of her or his behavior. Communication of feelings and immediacy is based on the interviewer's awareness of what is occurring, either personally or in the interview. Confrontation is based on the interviewer's effective listening and careful observation of the client's behavior.

The first half of this chapter acquaints you with the mechanics of confrontation. The second half illustrates situations in which confrontation is appropriate. The first sequence of frames in this program focuses on an interview between a counselor and a middle-aged woman whose marriage has failed. The client is seeking assistance because she is unable to manage her finances. In the latter part of the chapter, each frame is complete in itself. Go to 8.1.

8.1

When confronting, interviewers bring specific discrepancies to the attention of clients—discrepancies in their actions, in their words, and between their actions and words. Effective confrontation often leads the client to a clearer understanding of his or her problems and to constructive change.

CLIENT: (having indicated that she has just bought an expensive dress) I'm really worried about how I'm going to pay this month's rent.

Choose the most appropriate response.

INTERVIEWER: Not having enough money can be really stressful. (Go to 8.2)

INTERVIEWER: Maybe the bank will give you a loan. (Go to 8.3)

INTERVIEWER: I'm a bit confused. Earlier, you talked about buying an expensive dress, but now you say you're worried about paying your rent. It seems hard for you to manage your money. (Go to 8.4)

8.2   Your answer: *Not having enough money can be really stressful.*
Although your response accurately reflects the client's distress, you've failed to confront her questionable behavior. Your implied acceptance of the client's behavior may lead her to believe that her actions are appropriate. Return to 8.1 and try again.

8.3   Your answer: *Maybe the bank will give you a loan.*
Although a loan might provide a solution to the client's immediate financial difficulty, it won't help her deal with her underlying money-management problems. Moreover, you've failed to confront her with the money-management issue and have avoided discussing this relevant topic. Return to 8.1 and try again.

| **8.4** | **Your answer:** | *I'm a bit confused. Earlier, you talked about buying an expensive dress, but now you say you're worried about paying your rent. It seems hard for you to manage your money.* |

Correct. This response focuses tentatively and nonjudgmentally on a specific discrepancy in the client's communication. Go to 8.5.

8.5

The interviewer who uses confrontation effectively becomes a model for honest communication. Don't avoid using confrontation because you're concerned that the client will be hurt or because you're afraid of endangering your relationship with the client. On the other hand, don't use confrontation to express anger or to punish or get even.

CLIENT: It's really my former husband's fault. He refused to give me what I asked for when we divorced.

Choose the most appropriate response.

INTERVIEWER: You say your former husband has caused your financial problems, yet you spend money on expensive clothing. It's very hard for you to look at your role in this situation. (Go to 8.6)

INTERVIEWER: People who spend money the way you do and then blame someone else are just fooling themselves. (Go to 8.7)

INTERVIEWER: A fair settlement in a divorce is very important. Could you help me understand why your settlement was insufficient? (Go to 8.8)

8.6 **Your answer:** *You say your former husband has caused your financial problems, yet you spend money on expensive clothing. It's very hard for you to look at your role in this situation.*

Correct. By bringing the client's attention to the discrepancy between her behavior and her verbal message, you are helping her focus on the situation she is avoiding—her own role in her financial dilemma. Go to 8.9.

8.7 **Your answer:** *People who spend money the way you do and then blame someone else are just fooling themselves.*

This is an expression of feeling and opinion, not a confrontation. Moreover, this response won't help the client examine her behavior. Return to 8.5 and try again.

8.8 **Your answer:** *A fair settlement in a divorce is very important. Could you help me understand why your settlement was insufficient?*

You are in the midst of confronting the client about her inability to manage money. This response changes the focus and allows her to avoid her real problem. Return to 8.5 and try again.

8.9   *When confronted, the client must consider another way of acting, thus employing new ways of behaving. In using confrontation, you feed back discrepancies in behavior and communication and help the client explore those discrepancies.*

CLIENT: I need to get enough money to pay the rent. My unemployment check comes in next week, and I plan to buy a better DVD player with that.

Choose the most appropriate response.

INTERVIEWER: You told me you already have a DVD player. Why do you want to buy another one? (Go to 8.10)

INTERVIEWER: Shouldn't you spend your money on other things? (Go to 8.11)

INTERVIEWER: You're short of money, but you really find it difficult to stop spending. (Go to 8.12)

8.10   Your answer: *You told me you already have a DVD player. Why do you want to buy another one?*
This confrontation focuses on a minor discrepancy. Your response should direct the client's attention to her overall problem. Return to 8.9 and try again.

8.11   Your answer: *Shouldn't you spend your money on other things?*
This response may seem judgmental to the client and will probably cause her to feel defensive rather than open. Return to 8.9 and try again.

8.12   Your answer: *You're short of money, but you really find it difficult to stop spending.*
Correct. By confronting the client with her difficulty in a nonjudgmental manner, you ask her to consider her present behavior and the possibility of changing. Go to 8.13.

8.13   *Confrontation should be used only when a good relationship exists between the interviewer and the client. Good rapport is important because a confrontation frequently focuses on material that is stressful to the client. As a result, the client may become distressed or confused when confrontation is used.*

CLIENT: (tearfully) I know my debts are my responsibility, but it's so hard to admit it. Since my marriage broke up, I've just gone out and bought things without thinking. It has given me something to do.

Once she has faced her responsibility, the client, along with the interviewer, sets up a program to resolve her budgeting problems. The client is also concerned about obtaining a job to supplement her limited income.

CLIENT: (discussing a job for which she isn't qualified) That's another reason I want that job. I thought it would help me to meet people. I really want that job.

Choose the most appropriate response.

INTERVIEWER: That is a good job, isn't it? Perhaps another opening like that will come up soon. I'm sure that it would be good for you. (Go to 8.14)

INTERVIEWER: There seems to be some discrepancy between the type of job you look for and your actual qualifications. It seems hard for you to explore jobs that match your qualifications. (Go to 8.15)

INTERVIEWER: With your qualifications and experience, you shouldn't expect to get that sort of job. (Go to 8.16)

8.14 Your answer: *That is a good job, isn't it? Perhaps another opening like that will come up soon. I'm sure that it would be good for you.*
Your failure to confront the client with her inappropriate choice of jobs reinforces her behavior. She may be distressed by a confrontation, but unless you confront her, she will probably continue to seek unsuitable jobs. Return to 8.13 and try again.

8.15 Your answer: *There seems to be some discrepancy between the type of job you look for and your actual qualifications. It seems hard for you to explore jobs that match your qualifications.*
Correct. By focusing on the discrepancy between her qualifications and those required by the job, you help the client focus on an issue she has either avoided or failed to recognize. Go to 8.17.

8.16 Your answer: *With your qualifications and experience, you shouldn't expect to get that sort of job.*
This confrontation is far too abrupt and direct. Clients often feel threatened by confrontation. Your response should be helpful, not abusive or punitive. Return to 8.13 and try again.

8.17 *Clients are often more receptive to tentative, nonjudgmental confrontations. Tentative confrontations often begin with such phrases as could it be and you tend to suggest and continue in the same manner.*

CLIENT: Well, the job was just what I wanted. I'm not going to apply for something I don't want.

Choose the most appropriate response.

INTERVIEWER: You tend to suggest that the only thing you are considering is whether or not you want the job. It's difficult for you to think about other factors. (Go to 8.18)

INTERVIEWER: Could it be that yours is a shortsighted way to select a job? (Go to 8.19)

INTERVIEWER: Could it be that you just want your own way? (Go to 8.20)

8.18 Your answer: *You tend to suggest that the only thing you are considering is whether or not you want the job. It's difficult for you to think about other factors.*
Correct. This tentative confrontation will help the client explore her job-hunting behavior. Go to 8.21.

8.19 Your answer: *Could it be that yours is a shortsighted way to select a job?*
Although you began your response tentatively, you concluded by directly confronting the client with your own feelings and opinions about her behavior. Your response should invite the client to examine her behavior rather than react in her own defense. Return to 8.17 and try again.

8.20 **Your answer:** *Could it be that you just want your own way?*
Your tentative opening should be followed by an equally tentative confrontation. This blunt and direct response will be of no help to the client. Return to 8.17 and try again.

8.21 *Be aware of the client's feelings both during and after a confrontation. Good confrontations often include reflections of feeling concerning the client's difficulty in facing the discrepancy being discussed.*

CLIENT: That's all that's important. For years I've done what everybody else wanted me to do. Now it's time for me to do what I want to do.

Choose the most appropriate response.

INTERVIEWER: But what would happen if everyone acted that way? (Go to 8.22)

INTERVIEWER: It sounds as though your life has been absolute hell. Could you tell me about it? (Go to 8.23)

INTERVIEWER: Having been frustrated by a lack of choice in the past, you're now determined to do what you want to do. It's hard for you to realize that there is more to getting a job than just wanting it. (Go to 8.24)

8.22 **Your answer:** *But what would happen if everyone acted that way?*
This confrontation is far too direct. Moreover, it invites the client to discuss society in general rather than her problems in particular. Return to 8.21 and try again.

8.23 **Your answer:** *It sounds as though your life has been absolute hell. Could you tell me about it?*
By attending to the client's past, you are helping her avoid a difficult but important issue. Return to 8.21 and try again.

8.24 **Your answer:** *Having been frustrated by a lack of choice in the past, you're now determined to do what you want to do. It's hard for you to realize that there is more to getting a job than just wanting it.*
Correct. You have confronted the client while attending to her feelings concerning the problem area. By confronting with support, you invite the client to face her problem and deal with it. Go to 8.25.

8.25 *Base your confrontation on the client's statements or behavior, not on inferred information. Confrontations should not include accusations, evaluations, or solutions to problems.*

CLIENT: Yes. I guess I'm afraid of being pushed around again. I don't know why they worry about my qualifications. Isn't it important that I want to do the job?

Choose the most appropriate response.

INTERVIEWER:	You've probably thought that, because you're not really qualified, you don't have to take responsibility for getting the job. (Go to 8.26)
INTERVIEWER:	You're finding it difficult to face the fact that your qualifications interfere with your choice of jobs. Considering your own qualifications is threatening. (Go to 8.27)
INTERVIEWER:	You seem to be forgetting that people have to work hard to get their qualifications. (Go to 8.28)

8.26 **Your answer:** *You've probably thought that, because you're not really qualified, you don't have to take responsibility for getting the job.*
You have no information on which to base this statement. Base your confrontation on what the client has said or done. Return to 8.25 and try again.

8.27 **Your answer:** *You're finding it difficult to face the fact that your qualifications interfere with your choice of jobs. Considering your own qualifications is threatening.*
Correct. By focusing on the discrepancy between the type of job the client wants and the type of job for which she is qualified, you've helped her look at her situation realistically. Go to 8.29.

8.28 **Your answer:** *You seem to be forgetting that people have to work hard to get their qualifications.*
This response will not help the client explore her qualifications or her behavior. Your confrontation should focus on a discrepancy in the client's message. Return to 8.25 and try again.

8.29

The client may try to discredit you, argue with you, or pretend to agree with you when you use confrontation. When this happens, continue to confront the client.

| CLIENT: | You have a good job . . . and I'm as much entitled to one as you are. You just don't want me to have a good job. |

Choose the most appropriate response.

INTERVIEWER:	You seem to be angry with me because I've discussed your qualifications. It's easier to avoid that issue than it is to examine it. (Go to 8.30)
INTERVIEWER:	I worked hard for my qualifications. You'll have to work hard for yours. (Go to 8.31)
INTERVIEWER:	Of course I want you to have a good job. I'll help you look for one. (Go to 8.32)

8.30 **Your answer:** *You seem to be angry with me because I've discussed your qualifications. It's easier to avoid that issue than it is to examine it.*
Correct. You've continued to confront the client while inviting her to examine and discuss her behavior in terms of her feelings. This method of confronting is supportive. Go to 8.33.

8.31 **Your answer:** *I worked hard for my qualifications. You'll have to work hard for yours.*
Your response gives the client an opportunity to change topics. You haven't helped her examine her own behavior. Return to 8.29 and try again.

8.32	**Your answer:**	*Of course I want you to have a good job. I'll help you look for one.*

This response is not a confrontation. Although you may have assuaged the client's anger, you haven't helped her confront and examine her behavior. Return to 8.29 and try again.

8.33

The use of confrontation is often required in crisis situations in which the client's behavior is threatening to his or her welfare or the welfare of others.

CLIENT: What's the point? It's no use. . . . It's hopeless. I'll never get what I want. Why should I even go on?

Choose the most appropriate response.

INTERVIEWER: It all seems really difficult for you right now. Finding a job seems to be an impossible task. (Go to 8.34)

INTERVIEWER: You talk positively about finding a good job, but you say that it's all hopeless. Sometimes it's hard to keep trying. (Go to 8.35)

INTERVIEWER: I'm really concerned about you. It sounds as though you might do something drastic. (Go to 8.36)

8.34	**Your answer:**	*It all seems really difficult for you right now. Finding a job seems to be an impossible task.*

This is a good reflection of feeling, but in order to explore quickly the gravity of her feelings, you should confront the client about the meaning of her previous comments. Return to 8.33 and try again.

8.35	**Your answer:**	*You talk positively about finding a good job, but you say that it's all hopeless. Sometimes it's hard to keep trying.*

Correct. This supportive confrontation is the most effective way to explore the gravity of the client's feelings quickly. Go to 8.37.

8.36	**Your answer:**	*I'm really concerned about you. It sounds as though you might do something drastic.*

Your confrontation is too direct and threatening. It could cause the client to feel even more depressed. Confrontation should focus on discrepancies and lead to positive change. Return to 8.33 and try again.

8.37

Confrontation can be directed at discrepancies related to the client's strengths and resources as well as his or her weaknesses and limitations.

CLIENT: It's just that . . . I know I have the ability to retrain and get the sort of job I want . . . but I guess I find it easier to stay home and collect money from my former husband.

Choose the most appropriate response.

INTERVIEWER: You know you have the ability to earn money to buy the things you want, but something seems to make it hard for you to do that. (Go to 8.38)

INTERVIEWER: Don't you think you should try to change your attitude? With your ability, you could be out earning money for yourself. (Go to 8.39)

INTERVIEWER: Don't you think you're trying to get even with your former husband? (Go to 8.40)

8.38   **Your answer:**   *You know you have the ability to earn money to buy the things you want, but something seems to make it hard for you to do that.*

Correct. This confrontation focuses on an ability that the client isn't using at this time. By confronting the client in this way, you encourage her to explore her strengths and take some positive action. Go to 8.41.

8.39   **Your answer:**   *Don't you think you should try to change your attitude? With your ability, you could be out earning money for yourself.*

This confrontation is too blunt; it may provoke defensive or aggressive behavior on the part of the client. Return to 8.37 and try again.

8.40   **Your answer:**   *Don't you think you're trying to get even with your former husband?*

This confrontation implies something that may or may not be true. Confrontations must be based on information, not on assumptions. Return to 8.37 and try again.

8.41

It is important for an interviewer to know when confrontation is appropriate. The following frames offer examples of situations in which confrontation should be employed. Each frame represents a separate interview.

CLIENT: (whose records indicate that his smoking has increased slightly during the past 2 weeks) I know smoking is bad for my health; I'm cutting down.

Choose the most appropriate response.

INTERVIEWER: If you want to stop smoking, you'll have to try harder. (Go to 8.42)

INTERVIEWER: It's good to hear you say that you're cutting down. You sound committed to that. (Go to 8.43)

INTERVIEWER: You say you've cut down, but your records show that your smoking has increased. (Go to 8.44)

8.42   **Your answer:**   *If you want to stop smoking, you'll have to try harder.*

The client has probably heard this advice many times. Confront him about the discrepancy between his stated goals and his behavior; help him resolve this discrepancy and take responsibility for reducing his smoking. Return to 8.41 and try again.

8.43   **Your answer:**   *It's good to hear you say that you're cutting down. You sound committed to that.*

Although it's good to respond positively to the client's efforts, you've ignored the discrepancy between his words and his actions. In order to help him, you must confront the client with this discrepancy. Return to 8.41 and try again.

8.44   Your answer: *You say you've cut down, but your records show that your smoking has increased.*
Correct. By confronting the client about the discrepancy between his comments and his actions, you will help him resolve the discrepancy and take responsibility for his behavior. Go to 8.45.

8.45   Interviewers often note a discrepancy between clients' self-perception and their actual appearance.

CLIENT: (whose looks are above average) I just can't stand myself. I'm so ugly.

Choose the most appropriate response.

INTERVIEWER: You know, you're quite wrong. You really are very attractive. I'm sure you've been told that before. (Go to 8.46)

INTERVIEWER: You say you're ugly, but I'm impressed by how attractive you are. I wonder if you could sort out this contradiction. (Go to 8.47)

INTERVIEWER: I think you've been watching too much television. We don't all have to be sex symbols, you know. (Go to 8.48)

8.46   Your answer: *You know, you're quite wrong. You really are very attractive. I'm sure you've been told that before.*
You are probably correct in assuming that the client has been told she is attractive, so it's unlikely that you will help her by telling her this again. Direct her attention to the discrepancy between what she sees and what you see. Return to 8.45 and try again.

8.47   Your answer: *You say you're ugly, but I'm impressed by how attractive you are. I wonder if you could sort out this contradiction.*
Correct. This confrontation focuses the client's attention on the discrepancy between what she sees and what you see and asks her to examine it. Go to 8.49.

8.48   Your answer: *I think you've been watching too much television. We don't all have to be sex symbols, you know.*
You've discounted the client's feelings and made it difficult for her to examine her perception of herself. Confront her with the discrepancy between her perception and yours. Return to 8.45 and try again.

8.49   Confrontation can be used to draw a client's attention to incongruous or contradictory comments and behaviors.

CLIENT: (who hesitantly asserts that there are no problems in his marriage) My wife's going away again this weekend. As I've told you before, I'm really lucky to have such a good marriage.

Choose the most appropriate response.

INTERVIEWER: You feel you have a good marriage. (Go to 8.50)

INTERVIEWER: In my experience, good marriages don't involve separations every weekend. (Go to 8.51)

INTERVIEWER: You maintain that you have a good marriage, but you don't seem to be comfortable with your wife's absence on weekends. (Go to 8.52)

8.50 **Your answer:** *You feel you have a good marriage.*
This is a good reflection of feeling, but you should confront the client about the incongruity between his wife's behavior and his apparent belief that all is well with his marriage. Return to 8.49 and try again.

8.51 **Your answer:** *In my experience, good marriages don't involve separations every weekend.*
This confrontation is likely to produce defensive behavior on the part of the client. Your confrontation should be supportive and helpful. Return to 8.49 and try again.

8.52 **Your answer:** *You maintain that you have a good marriage, but you don't seem to be comfortable with your wife's absence on weekends.*
Correct. Your response confronts the client with the discrepancy between his statements about his marriage and his feelings about his wife's behavior. Go to 8.53.

8.53 Confrontation can be used to draw clients' attention to discrepancies between what they say and how they express themselves.

CLIENT: (with head down; in low, flat, expressionless voice) I'm really very happy. I don't know why I'm sitting here. Everything is going so well.

Choose the most appropriate response.

INTERVIEWER: You say you're happy, but your expression and tone of voice suggest otherwise. (Go to 8.54)

INTERVIEWER: It's really good to hear you say everything is going so well. People don't often feel that way. (Go to 8.55)

INTERVIEWER: I just can't believe that. How can you sit there looking depressed and tell me that you're happy? (Go to 8.56)

8.54 **Your answer:** *You say you're happy, but your expression and tone of voice suggest otherwise.*
Correct. By focusing on the discrepancy between the client's statements and her way of expressing them, you've confronted her with an inconsistency and encouraged her to explore it. Go to 8.57.

8.55 **Your answer:** *It's really good to hear you say everything is going so well. People don't often feel that way.*
You've ignored a large portion of the client's total message. She needs feedback about her behavior if she is to confront any difficulties she is avoiding. Return to 8.53 and try again.

8.56   **Your answer:**   *I just can't believe that. How can you sit there looking depressed and tell me that you're happy?*
This confrontation is too abrupt and will probably cause the client to become defensive. Your confrontation should encourage the client to explore her behavior, not defend it. Return to 8.53 and try again.

8.57   *Rather than attempt to change their behavior, clients often find excuses, blame others, forget a problem, or simply apologize for their behavior. Confrontation can draw these defensive strategies to the clients' attention.*

CLIENT: (overweight and trying to diet) I don't see why I can't eat whatever I want to eat. Everybody else does. I like to eat.

Choose the most appropriate response.

INTERVIEWER: You must realize that it's important for you to lose weight. The sooner you get on with it, the better. (Go to 8.58)

INTERVIEWER: Could it be that eating is more important to you than losing weight? (Go to 8.59)

INTERVIEWER: It seems that you really don't want to do anything about your weight problem. (Go to 8.60)

8.58   **Your answer:**   *You must realize that it's important for you to lose weight. The sooner you get on with it, the better.*
Since the client is defending his behavior, it is unlikely that he will heed your advice. Draw the client's attention to his behavior and encourage him to deal with it. Return to 8.57 and try again.

8.59   **Your answer:**   *Could it be that eating is more important to you than losing weight?*
Correct. This confrontation focuses the client's attention on his behavior and invites him to explore it. Go to 8.61.

8.60   **Your answer:**   *It seems that you really don't want to do anything about your weight problem.*
This expression of your own opinion will probably cause the client to become defensive. Return to 8.57 and try again.

8.61   *It's often necessary to confront clients with contradictions in their personal and social values.*

CLIENT: I believe women have the right to work and contribute to society. I work with some very competent women. But I'm happy that my wife stays home to help me entertain clients. I could never let her work. . . . How would it look?

Choose the most appropriate response.

INTERVIEWER: You think it's very important that your wife supports you in your business activities. (Go to 8.62)

INTERVIEWER: Isn't your attitude toward your wife a bit chauvinistic? (Go to 8.63)

INTERVIEWER: You say you think women have the right to work, but you find it hard to extend this attitude to include your wife. (Go to 8.64)

8.62 **Your answer:** *You think it's very important that your wife supports you in your business activities.* You should confront the client with the contradictory statements he has made so that he can examine his attitudes. Return to 8.61 and try again.

8.63 **Your answer:** *Isn't your attitude toward your wife a bit chauvinistic?* This response will cause the client to become defensive or hostile. Confront the client with the discrepancy between his personal and social values and help him resolve this discrepancy. Return to 8.61 and try again.

8.64 **Your answer:** *You say you think women have the right to work, but you find it hard to extend this attitude to include your wife.* Correct. This response confronts the client with the discrepancy between his personal and social values. Go to 8.65.

8.65 *Confront clients about their unrealistic goals to help them formulate workable goals.*

CLIENT: (who is on a fixed income and is in debt) I've been looking at houses. I'm sure that, if we had a house, the children would settle down.

Choose the most appropriate response.

INTERVIEWER: It's important to you to own a house, but I wonder how realistic that is for someone in your financial situation. (Go to 8.66)

INTERVIEWER: It's important that you give your children the security of a home. You've identified a goal for yourself. (Go to 8.67)

INTERVIEWER: I'm not sure that's a good idea. Your children are more likely to improve if you settle down and stop creating debts. (Go to 8.68)

8.66 **Your answer:** *It's important to you to own a house, but I wonder how realistic that is for someone in your financial situation.* Correct. This confrontation encourages the client to examine his aspirations and set realistic goals for himself. Go to 8.69.

8.67 **Your answer:** *It's important that you give your children the security of a home. You've identified a goal for yourself.* If you don't confront the client, he will assume that you support his unrealistic goals. Return to 8.65 and try again.

8.68 **Your answer:** *I'm not sure that's a good idea. Your children are more likely to improve if you settle down and stop creating debts.* Confrontation, rather than advice, will enable the client to examine his goals and change them. Return to 8.65 and try again.

8.69	*Because of its strong effect on clients, always use confrontation in context and never use it excessively.*

CLIENT: I hope it's all right with you if we talk today about a problem that's caused me a lot of anxiety this week.

Choose the most appropriate response.

INTERVIEWER: Last week we agreed to continue discussing an important area, but now you want to discuss something else. I get the feeling that you're avoiding the problem we were talking about last week. (Go to 8.70)

INTERVIEWER: That's fine. It sounds like a matter of real concern to you. Start wherever you want to start. (Go to 8.71)

INTERVIEWER: I think we should continue to work on the area we were discussing last week. (Go to 8.72)

8.70   Your answer: *Last week we agreed to continue discussing an important area, but now you want to discuss something else. I get the feeling that you're avoiding the problem we were talking about last week.*

This confrontation would be appropriate if you were able to evaluate the importance of the problem the client wants to discuss. The client's behavior seems appropriate at this time; a confrontation isn't needed. Return to 8.69 and try again.

8.71   Your answer: *That's fine. It sounds like a matter of real concern to you. Start wherever you want to start.*

Correct. No confrontation is required at this time. You've encouraged the client to discuss the source of his new difficulties. Go to 8.73.

8.72   Your answer: *I think we should continue to work on the area we were discussing last week.*

It is inappropriate for you to disagree with the client at this point. Listen to what he has to say and, if necessary, confront him when you have information that warrants using confrontation. Return to 8.69 and try again.

8.73	# Review Questions

Indicate whether each of the following statements is true or false.

1. Confrontation draws the client's attention to specific discrepancies in behavior.
2. Whether or not the interviewer confronts a client is determined by his or her feelings about the client.
3. Rapport between the interviewer and the client is irrelevant to the appropriateness of confrontation.
4. Tentative rather than direct confrontations are likely to have an effect on the client.

5. The client's feelings should be respected both during and after a confrontation.
6. Confrontations often include evaluative comments or solutions to problems.
7. Confrontations should be used only when dealing with the client's weaknesses or limitations.
8. Interviewers should not confront clients about discrepancies between their own perceptions and the clients'.

Review Answers

1. *True.* If you answered *false,* go to 8.1 and review.
2. *False.* If you answered *true,* go to 8.5 and review.
3. *False.* If you answered *true,* go to 8.13 and review.
4. *True.* If you answered *false,* go to 8.17 and review.
5. *True.* If you answered *false,* go to 8.21 and review.
6. *False.* If you answered *true,* go to 8.25 and review.
7. *False.* If you answered *true,* go to 8.37 and review.
8. *False.* If you answered *true,* go to 8.45 and review.

If six or more of your responses were incorrect, you should return to 8.1 and review this chapter.

Points to Remember about Confronting

1. WHEN IDENTIFYING DISCREPANCIES:
   a. Focus on observed discrepancies.
   b. Focus on discrepancies that are related to clients' strengths as well as their limitations.
2. WHEN YOU CONFRONT:
   a. State the discrepant elements in the client's message and encourage the client to explore these discrepancies.
   b. Be tentative.
   c. Be prepared to explore feelings.
   d. Don't use this skill as a means of punishment or revenge.
   e. Your comments shouldn't include accusations, judgments, or solutions to problems.
3. USE CONFRONTATION:
   a. To show the client how to recognize contradictions and resolve them.
   b. To help the client identify and resolve discrepancies.
   c. After you've established a good relationship with a client.
   d. Whenever a client's behavior is a threat to his or her welfare or the welfare of others.
   e. When you recognize discrepancies between the client's words and actions, between the client's perceptions and your own, between the

client's message and the way that message is communicated, or between the client's personal and social values.

f. When the client exhibits incongruous or contradictory behavior patterns.
g. When the client employs defensive strategies.
h. When the client sets unrealistic goals.

8.1

Activity Unit

Find two others who have also read this chapter. Role play short interviews alternating the client, interviewer, and observer roles. Record each brief interview on audio- or videotape. An interview with two or three confronting responses and other associated responses necessary to resolve each confrontation is sufficient for this exercise.

Interviewer role Practice using confronting responses and interviewing responses required to complete the confrontation.

Client role Be cooperative and facilitate the progress of the interview so the interviewer can practice the skill of confronting. Clients may want to role play the problem of a friend or family member that includes discrepant communication or to enact one of the following roles:

1. A spouse who is always working and has indicated he or she will pay more attention to the family. During the interview, the week's activities are discussed, and no change has occurred.
2. An individual who has just achieved something important but continually negates that success.
3. A person whose comments indicate considerable distress about a child's leaving home but who angrily professes a lack of interest in the child's welfare.
4. An employee who has difficulty cooperating with fellow workers, who sits with arms tightly folded, and who avoids eye contact and discussion in response to the personnel officer's attempts to discuss the matter.

Observer role Operate the equipment and complete the Practice Interview Checklist (see Table 8.1).

Postinterview discussion Using the taped interviews and the Practice Interview Checklist data, the participants should discuss the performance of the interviewer. All participants can learn from this discussion, which should be nonjudgmental, focus on positive as well as less effective responses, and stimulate improved performance for all.

Table 8.1

Practice Interview Checklist

Activity	Interviewer response[a]				
	1	2	3	4	5
Appropriate nonverbal behavior					
On topic					
Confronting response					
Other interviewing response					
Effective response[b]					

[a] Check the activities that characterize each response.

[b] Indicate the degree of effectiveness on a scale from 1 to 5 in which 1 is ineffective and 5 is very effective.

8.2

Activity Unit

Using the tape of the interview produced in Activity Unit 8.1 or the tape of another practice interview, enter below up to three of the original interview confrontations below. Then formulate a better confrontation to replace each original one in the space provided.

Original response 1 _____

_____

_____

Better response _____

_____

_____

Original response 2 _____

_____

_____

Better response _____

_____

_____

Original response 3 _____

_____

_____

Better response _____

_____

_____

Cultural Considerations

Clients from some cultural groups including Native North Americans, Canadian Inuit, and traditional Latino/Latina people may view confrontation as disrespectful and insensitive. With clients from these groups, interviewers should choose alternate culturally appropriate skills.

InfoTrac® College Edition Keyword Search Terms

Confronting AND Counseling

Chapter Nine

Self-Disclosing

*T*his chapter is intended to help you master the skill of self-disclosing. After completing this chapter, you should be able to:

1. Self-disclose to a client.
2. Explain the purpose of self-disclosing.
3. Decide whether self-disclosure is appropriate in the situation.

Self-disclosure occurs when an interviewer intentionally reveals personal information to a client in the hope of assisting the client. Self-disclosure is used to facilitate clients' understanding of their experiences, thoughts, feelings, and behaviors, to provide information, and to help clients to develop strategies and resolve issues. Interviewer self-disclosure facilitates client self-disclosure, creates additional trust, and can instill hope in a client who feels alone and without support.

There has been considerable discussion about the use of self-disclosing as an interviewing skill. Humanistically oriented and some interpersonal interviewers have presented evidence to substantiate the facilitative effects of an interviewer's self-disclosure on the interviewing process. In contrast, some interviewers maintain that self-disclosure on the part of the interviewer can change the focus of the discussion from the client's concerns to those of the interviewer. When this occurs, the natural flow of the interview is disturbed. Interviewers holding this position argue that the interviewer is most effective when objectivity is maintained and self-disclosures are avoided.

Although interviewer self-disclosure occurs infrequently in most interviews, research findings generally indicate that clients and interviewers find the skill helpful and important. Improved quality and quantity of client discussion, increased trust, and the establishment of a more balanced relationship have been observed to result from the appropriate use of interviewer self-disclosures. The skill has been identified as having a particular importance in self-help groups. It is important for you to learn both how and when to employ the skill of self-disclosure if you intend to use it.

From this chapter, you will learn how to formulate self-disclosures, how self-disclosures affect clients, and when it is appropriate to use self-disclosure. In this chapter, one self-disclosure follows another to help you develop this skill. In an actual interview, however, the interviewer should not use one self-disclosure after another.

The frames in this chapter follow an interview between a high school counselor and a student. The counselor has helped the student in the past, and they have developed a good relationship. This interview was requested by the student. Go to 9.1.

9.1

Self-disclosure means sharing personal information about yourself, your experiences, your attitudes, and your feelings. A good interviewer discloses such information if it appears that it will help the client.

CLIENT: I'm really upset. I just found out that my brother and sister are using drugs.

Choose the most appropriate response.

INTERVIEWER: How can you be sure they're on drugs? (Go to 9.2)

INTERVIEWER: I'm glad you found out about it. I can understand your concern. (Go to 9.3)

INTERVIEWER: You must be worried. When my brother was on drugs, I was really worried about him. I can understand your concern. (Go to 9.4)

9.2   Your answer: *How can you be sure they're on drugs?*

This question could elicit important information; however, in this chapter, you are learning to self-disclose. A self-disclosure involves telling the client something personal about yourself that relates to his or her problem. Return to 9.1 and try again.

9.3	**Your answer:**	*I'm glad you found out about it. I can understand your concern.* This self-disclosure reveals very little about you. Return to 9.1 and try again.

9.4	**Your answer:**	*You must be worried. When my brother was on drugs, I was really worried about him. I can understand your concern.* Correct. You've revealed personal information that lets the client know you understand his feelings. The client will feel comfortable discussing the problem with you. Go to 9.5.

9.5

Self-disclosure involves (1) using such pronouns as I *or* me *and (2) reference to some personal feeling or event that is relevant to the client's discussion.*

CLIENT: Yes, I'm worried about them. And I don't know what to do about it. As you know, my father died 2 years ago. My mother would get upset if I told her about my brother and sister. I have to do something.

Choose the most appropriate response.

INTERVIEWER: It must be difficult for you. Why don't you talk to a relative about it? I would. (Go to 9.6)

INTERVIEWER: It must be difficult for you. It's important for me to talk over problems with my family. I feel alone until I do. (Go to 9.7)

INTERVIEWER: It must be difficult for you. I really think that you should tell your mother. She'll have to know eventually. (Go to 9.8)

9.6	**Your answer:**	*It must be difficult for you. Why don't you talk to a relative about it? I would.* This reflection of feeling, and the leading question that follows it, will be of little use to the client. You could use appropriate self-disclosure to help the client with his problem. Return to 9.5 and try again.

9.7	**Your answer:**	*It must be difficult for you. It's important for me to talk over problems with my family. I feel alone until I do.* Correct. This is an appropriate self-disclosure that shows the client that you understand the situation. Go to 9.9.

9.8	**Your answer:**	*It must be difficult for you. I really think that you should tell your mother. She'll have to know eventually.* This reflection of feeling, followed by what could be inappropriate advice, won't help the client solve the problem he has presented. Return to 9.5 and try again.

9.9

The timing of an interviewer's self-disclosure is important. Premature self-disclosures disrupt the interviewing process. Well-timed and well-focused self-disclosures enable the client to go on with the interview without distraction.

CLIENT: I'm planning to go away to college in the fall. If I have to worry about this problem, I don't see how I can go.

Choose the most appropriate response.

INTERVIEWER: That certainly complicates things. I have the same kind of problem at the moment. I may have to move soon, and I will have to consider my family. (Go to 9.10)

INTERVIEWER: You must feel considerable pressure. When I'm faced with a problem like yours, I try to identify its central aspect. It's hard to face the whole thing at once. (Go to 9.11)

INTERVIEWER: I find that as soon as one problem crops up, it leads to a string of others. I wish someone would come up with an explanation of why that happens. (Go to 9.12)

9.10   Your answer: *That certainly complicates things. I have the same kind of problem at the moment. I may have to move soon, and I will have to consider my family.*
This irrelevant self-disclosure will distract the client. As the interviewer, your role is to facilitate discussion, not divert it. Return to 9.9 and try again.

9.11   Your answer: *You must feel considerable pressure. When I'm faced with a problem like yours, I try to identify its central aspect. It's hard to face the whole thing at once.*
Correct. This timely self-disclosure is on topic and will help the client explore his problem. Go to 9.13.

9.12   Your answer: *I find that as soon as one problem crops up, it leads to a string of others. I wish someone would come up with an explanation of why that happens.*
This kind of irrelevant self-disclosure has no place in an interview. Return to 9.9 and try again.

9.13

When using self-disclosure, you can share information that relates to your experiences, your current situation, or possible events in your future. Self-disclosures that relate to your current situation are most powerful because they help the client focus on a particular problem and deal with it concretely.

CLIENT: I am a little overwhelmed. I suppose I just have to do something about my brother and sister. If I don't do something about them, I'll have a hard time going off to college. I don't know how they would manage. . . . I guess I should tell my mother and let her deal with it, but I just don't know how she would cope.

Choose the most appropriate response.

INTERVIEWER: At times, I feel responsible for one or both of my parents. When that happens, I often underestimate how well they can cope. When I discuss problems with them, I'm surprised at how well they can deal with them. (Go to 9.14)

INTERVIEWER: When my father dies, I know I'll have to assume responsibility in the same way you have. And that will create problems in my life. I guess I'll try to find solutions to problems on my own. (Go to 9.15)

INTERVIEWER: When I was your age, I often found it hard to discuss things with my mother. I tried to solve problems on my own, but later I wished I had asked her for help. (Go to 9.16)

9.14 **Your answer:** *At times, I feel responsible for one or both of my parents. When that happens, I often underestimate how well they can cope. When I discuss problems with them, I'm surprised at how well they can deal with them.*

Correct. This self-disclosure focuses on the present situation and helps the client identify an important aspect of his problem. Go to 9.17.

9.15 **Your answer:** *When my father dies, I know I'll have to assume responsibility in the same way you have. And that will create problems in my life. I guess I'll try to find solutions to problems on my own.*

This response indicates that you don't understand or empathize with the client's situation. Your response should encourage the client, not inhibit him. Return to 9.13 and try again.

9.16 **Your answer:** *When I was your age, I often found it hard to discuss things with my mother. I tried to solve problems on my own, but later I wished I had asked her for help.*

This self-disclosure is appropriate, but it isn't as powerful as one that focuses on the present. Return to 9.13 and try again.

9.17

Self-disclosure allows the client to know the interviewer as a person, facilitating a greater level of trust and open discussion between the two.

CLIENT: I don't really think my mother can cope with my brother's and sister's problem. I don't really know, though. I haven't let her handle very many responsibilities since my father died. Maybe I should tell my mother about all of this, but I've always found it hard to talk to her. I don't really know how to go about it.

Choose the most appropriate response.

INTERVIEWER: I can't talk to my mother, either. She's always busy. (Go to 9.18)

INTERVIEWER: You know, my mother's funny. When I think she's listening, she doesn't hear a thing. (Go to 9.19)

INTERVIEWER: I find it hard to talk to my mother, too. It's all right once I'm started, but it's always difficult for me at the beginning. (Go to 9.20)

9.18 **Your answer:** *I can't talk to my mother, either. She's always busy.*

There is little in this self-disclosure that is relevant to the client's problem. Your response will distract him from considering the feelings he has about communicating with his mother. Return to 9.17 and try again.

9.19 **Your answer:** *You know, my mother's funny. When I think she's listening, she doesn't hear a thing.*

This self-disclosure doesn't necessarily relate to the client's problem. The client has told you that he has difficulty discussing things with his mother, but you don't know anything about his mother's listening abilities. To build mutual trust, you must share personal information that relates to the topic under discussion. Return to 9.17 and try again.

9.20   **Your answer:**   *I find it hard to talk to my mother, too. It's all right once I'm started, but it's always difficult for me at the beginning.*
Correct. This self-disclosure focuses on your reaction to a situation similar to the client's. By revealing this information, you help develop mutual trust and stimulate relevant discussion. Go to 9.21.

9.21   *Clients often find it difficult to share personal information. They may be afraid that sharing such information is inappropriate, they may be uncomfortable with self-disclosure, or they may not know how to self-disclose. Under these circumstances, the interviewer should model the skill of self-disclosing for the client.*

CLIENT: I really should discuss my concerns with my mother. I've known that for ages. I guess it's just like the other times I've had to discuss things with her. I keep putting it off, thinking about it . . . worrying about it . . . and then when I talk to her, I wonder why I waited so long. I'll tell her when I go home. It won't be easy. But I need to consider my plans for college.

Choose the most appropriate response.

INTERVIEWER: I often find that worrying about something is more stressful than actually doing it. And worrying keeps me from doing other things. (Go to 9.22)

INTERVIEWER: Thinking about things beforehand is really a problem for me. I don't know what I can do about it. (Go to 9.23)

INTERVIEWER: I've been thinking about college as well, but I can't decide whether to go back. It's a difficult decision. (Go to 9.24)

9.22   **Your answer:**   *I often find that worrying about something is more stressful than actually doing it. And worrying keeps me from doing other things.*
Correct. Your self-disclosure invites the client to stay on topic and consider the relevant aspects of his problem. Hence, your self-disclosure is an appropriate model for the client. Go to 9.25.

9.23   **Your answer:**   *Thinking about things beforehand is really a problem for me. I don't know what I can do about it.*
This self-disclosure changes the topic to one of your problems and could easily lead to a discussion of your situation rather than the client's. Self-disclosures should be on topic and advance the interview. Return to 9.21 and try again.

9.24   **Your answer:**   *I've been thinking about college as well, but I can't decide whether to go back. It's a difficult decision.*
Your self-disclosure changes the focus of the interview from the client to you. You've changed the topic and burdened the client with your problem. Return to 9.21 and try again.

9.25

Self-disclosure should help the client focus clearly and accurately on a particular problem and on the resources that can be used to solve that problem.

CLIENT: You know, when I came in, I thought my problem with my brother and sister might prevent me from going to college. But now, as I'm talking about it, I don't think it's just that. I think I'm afraid to leave home.

Choose the most appropriate response.

INTERVIEWER: When I went to college, I think I found it difficult to leave home, too. (Go to 9.26)

INTERVIEWER: I sometimes focus on the wrong reason for a problem, too. I often find that, after I solve one problem, I can move on to attend to others. (Go to 9.27)

INTERVIEWER: When I think about moving to a new position in a new city, I get uptight, too. (Go to 9.28)

9.26   Your answer: *When I went to college, I think I found it difficult to leave home, too.*
This self-disclosure is too general to help the client focus accurately on a particular problem or resource. Return to 9.25 and try again.

9.27   Your answer: *I sometimes focus on the wrong reason for a problem, too. I often find that, after I solve one problem, I can move on to attend to others.*
Correct. This self-disclosure relates to the situation that the client is describing. You've also offered a method of approaching his situation. Go to 9.29.

9.28   Your answer: *When I think about moving to a new position in a new city, I get uptight, too.*
This self-disclosure is off topic and provides no useful information to the client. Return to 9.25 and try again.

9.29

An interviewer must be able to rationalize that using self-disclosure at that time will benefit the client. An interviewer's self-disclosures should be concise and should not overshadow, deny, or put down what the client has said.

CLIENT: I really am afraid to leave home and go to college. I have a lot of difficulty making new friends. I'll be so lonely there. I suppose it really would help if I knew how to go about making new friends.

Choose the most appropriate response.

INTERVIEWER: I find that I'm better off at home rather than out trying to find new friends. (Go to 9.30)

INTERVIEWER: Trying to make new friends and failing always makes me feel worse. I've found that it's better to stick with old friends. (Go to 9.31)

INTERVIEWER: Sorting out things ahead of time really helps me. (Go to 9.32)

9.30   Your answer: *I find that I'm better off at home rather than out trying to find new friends.*
Your self-disclosure discourages the client from exploring possible solutions to his dilemma. Return to 9.29 and try again.

9.31   Your answer: *Trying to make new friends and failing always makes me feel worse. I've found that it's better to stick with old friends.*

In his previous statement, the client suggested a positive move that could reduce much of his anxiety about going away to college. This self-disclosure discounts that solution. Return to 9.29 and try again.

9.32   Your answer: *Sorting out things ahead of time really helps me.*

Correct. Your self-disclosure encourages the client to explore appropriate ways to work through his anxiety. Go to 9.33.

9.33   *Self-disclosing has been shown to be an effective interviewing technique when used appropriately. However, extensive use of interviewer self-disclosure can be problematic. Interviewers who use self-disclosing too frequently risk being viewed by clients as in need of assistance themselves, with a consequent negative effect on the client's confidence in the interviewer. When using this skill, the interviewer must keep the focus on the client and the client's needs. Go to 9.34.*

9.34

Review Questions

Indicate whether each of the following statements is true or false.

1. Self-disclosure can be used to demonstrate that it is appropriate to reveal personal information during an interview.
2. Self-disclosures are appropriate at any time in an interview.
3. Self-disclosure invites the client to share personal information with the interviewer.
4. Self-disclosure should be used to distract the client from specific problems.
5. Self-disclosure should be used as frequently as possible.
6. Self-disclosure generally facilitates trust between the interviewer and the client.
7. Self-disclosure is often used to help the client focus on a specific problem.
8. Self-disclosures that focus on past experiences are the most effective.

Review Answers

1. *True.* If you answered *false,* go to 9.21 and review.
2. *False.* If you answered *true,* go to 9.9 and review.
3. *True.* If you answered *false,* go to 9.21 and review.
4. *False.* If you answered *true,* go to 9.9 and review.
5. *False.* If you answered *true,* go to 9.33 and review.
6. *True.* If you answered *false,* go to 9.17 and review.
7. *True.* If you answered *false,* go to 9.25 and review.
8. *False.* If you answered *true,* go to 9.13 and review.

If five or more of your responses were incorrect, you should return to 9.1 and review the material in this chapter.

Points to Remember about Self-Disclosing

1. WHEN YOU SELF-DISCLOSE:
   a. Include personal information relevant to the client's situation.
   b. Focus on your present circumstances when possible.
   c. Be able to define the benefit of the response.
2. SELF-DISCLOSURE:
   a. Encourages the client to share information that is personally meaningful.
   b. Increases trust between the interviewer and the client.
   c. Enhances the client's ability to share feelings and personal information.
3. USE SELF-DISCLOSURE:
   a. To help the client focus clearly and accurately on problems and available resources.
   b. When your response won't overshadow, deny, or negate the client's communication.
   c. After you have established a good relationship with the client.
   d. In moderation.

9.1 Activity Unit

Working with two others who have also read this chapter, role play short interviews alternating the client, interviewer, and observer roles. Record each brief interview on audio- or videotape. An interview with five interviewer responses, including one or more self-disclosures, is sufficient for this exercise.

Interviewer role Practice self-disclosing in combination with other basic interviewing skills.

Client role Be cooperative and facilitate the progress of the interview so the interviewer can practice the skill of self-disclosing. Clients may want to role play the problem of a friend or family member or to enact one of the following roles:

1. A parent concerned about discipline and the risks of becoming abusive with a difficult-to-manage child.
2. A husband or wife concerned about an increase in the number of family arguments.
3. An employee having interpersonal difficulties with an immediate supervisor.
4. A person very upset over being charged with a minor traffic offense.

Table 9.1

Practice Interview Checklist

Activity	Interviewer response[a]				
	1	**2**	**3**	**4**	**5**
Appropriate nonverbal behavior					
On topic					
Self-disclosure					
Other appropriate interviewing response					
Effective response[b]					

[a] Check the activities that characterize each response.
[b] Indicate the degree of effectiveness on a scale from 1 to 5 in which 1 is ineffective and 5 is very effective.

Observer role Operate the equipment and complete the Practice Interview Checklist (see Table 9.1).

Postinterview discussion Using the taped interview and the Practice Interview Checklist data, the participants should discuss the performance of the interviewer. All participants can learn from this discussion, which should be nonjudgmental, focus on positive as well as less effective responses, and stimulate improved performance for all.

9.2 Activity Unit

Using the tape of an interview produced in Activity Unit 9.1 or the tape of another practice interview, enter the original interview responses below. Then formulate a better response to replace each original one in the space provided.

Original response 1 _____

_____

Better response _____

_____

_____

Original response 2 _____

_____

_____

Better response _____

_____

_____

Original response 3 _____

_____

_____

Better response _____

_____

_____

Original response 4 _____

_____

_____

Better response _____

_____

_____

Original response 5 _____

_____

_____

Better response _____

_____

_____

Cultural Considerations

1. Be aware that in some cultures self-disclosing is valued, whereas in others it is considered inappropriate. Those who have undergone traumatic experiences may find self-disclosure very difficult and prefer to avoid its use.
2. When using self-disclosing with a person of another cultural affiliation, inform yourself of the meaning and use of self-disclosing within the culture. Based on this knowledge and the progress of the interview, decide on the benefit of using and modeling the skill.

InfoTrac® College Edition Keyword Search Terms

Self-Disclosure AND Communication
Self-Disclosure AND Counseling
Self-Disclosure AND Interviewing

Chapter Ten

Information Giving

*T*his chapter is intended to help you master the skill of information giving. After completing this chapter, you should be able to:

1. Define each type of information-giving response.
2. Decide when an information-giving response is appropriate.
3. Formulate and use information-giving responses.

Regardless of an interviewer's theoretical background, the skill of information giving is important among the interviewer's range of responses. Information is given by interviewers in numerous fields, including health, education, community service, business, and industry. The ability to give information is essential during problem solving, decision making, and the provision of feedback.

Information can be given in a variety of ways. An interviewer may be called upon to provide information about the interviewing process, to give instructions and directions, to present feedback, or to provide or direct the client to

Table 10.1

An Overview of the Types of Information-Giving Responses

Type of response	Potential sources of information	Content of the response	Examples
Orienting statement	Program guidelines, procedure manual, policy manual, knowledge base	Information about a program, a process, or a procedure	Overview of the counseling process, introduction to a rehabilitation service, outline of a dental procedure
Instructions or directions	Instruction sheet, procedure manual, agency policy, knowledge base	Information describing a proposed behavior, how to carry it out, and its potential positive and negative effects	The client's role during the period of involvement, test instructions
Feedback	Test data, performance appraisal, consultant reports, observational data, laboratory reports	Information about performance, personal characteristics, or outcomes	Test results, performance review, health review
Alternative perspective (reframe)	Client's description of events, resources, people, alternatives, outcomes, procedures	Information about alternative frames of reference describing the situation	Half-full or half-empty glass
Informational statement	Knowledge base, reference source, data bank, library, educational materials	Information about events, resources, people, or alternatives	Overview of information-seeking skills, occupational information, description of community resources

knowledge relevant to the client's concern. To facilitate the client's goals, the interviewer requires both a sound knowledge base and well-developed communication skills. An overview of the types of information-giving responses and the potential sources of information associated with each is shown in Table 10.1.

From time to time, clients require varying amounts of information. Some interviews may be focused entirely upon mutual exchange of information within a facilitative atmosphere developed by the interviewer. Failure of the interviewer to provide information when required is counterproductive to the interviewing process. Until information is available, clients may be unable to take constructive action. It is important that interviewers recognize the needs of clients for information and provide this as required, either verbally or by directing clients to the facts.

The frames in this chapter relate to a series of interviews between a rehabilitation counselor and an employee who has been injured on the job. The task of the counselor is to assist the client to identify an alternate position within the company. Go to 10.1.

10.1

Information-giving responses involve the verbal communication by the interviewer to the client of data or facts about people, activities, events, resources, alternatives, outcomes, or procedures.

CLIENT: (having just arrived for the initial interview) I'm not sure whether it is you I should be asking, but I need to know what happens with my pay if I can't return to work.

Choose the most appropriate response.

INTERVIEWER: That's an important question, but we need to do other things before we get to it. (Go to 10.2)

INTERVIEWER: That information is in your company booklet on benefits, or at the end of this interview you can check with the payroll clerk, who can give you that information quickly. (Go to 10.3)

INTERVIEWER: I'm not sure about the reason you require this information. Can you tell me whether or not you are short of money, or is there some other problem? (Go to 10.4)

10.2   Your answer: *That's an important question, but we need to do other things before we get to it.*
The client has requested direct information. Failure of the interviewer to respond can result in client preoccupation with this question during the interview. Return to 10.1 and try again.

10.3   Your answer: *That information is in your company booklet on benefits, or at the end of this interview you can check with the payroll clerk, who can give you that information quickly.*
Correct. This response gives information about resources from which the client can obtain the information requested. Go to 10.5.

10.4   Your answer: *I'm not sure about the reason you require this information. Can you tell me whether or not you are short of money, or is there some other problem?*
Your request for information fails to focus on the client's response or to provide the client with the information required. Return to 10.1 and try again.

10.5

The interviewer's timing of information giving is critical. The information must match the client's current needs and goals, and must be provided at the point at which the client is most receptive.

CLIENT: I really want to work and, if everything goes smoothly, I hope I can find another job within the company. But I was wondering, if this doesn't happen, what is the possibility of a lump-sum settlement from the company?

Choose the most appropriate response.

INTERVIEWER: This company has very good lawyers. If you are thinking of a financial settlement, you should find a good attorney. (Go to 10.6)

INTERVIEWER: Given your concern about financial support, it sounds as if you should be seeking an appointment at Workers' Compensation. (Go to 10.7)

INTERVIEWER: Your goal and the goal of these sessions are similar: to find you another position within the company. It is too early to discuss any other outcome as yet. (Go to 10.8)

10.6   Your answer: *This company has very good lawyers. If you are thinking of a financial settlement, you should find a good attorney.*

Clients use only the information that fits their needs. Information about the legal aspects of the situation has not been solicited by the client, who may perceive such information as a threat rather than an attempt to resolve the problem. Return to 10.5 and try again.

10.7   Your answer: *Given your concern about financial support, it sounds as if you should be seeking an appointment at Workers' Compensation.*

The information contained in this advice is premature and possibly irrelevant to the client's concern. As such, it will most likely be ignored. Return to 10.5 and try again.

10.8   Your answer: *Your goal and the goal of these sessions are similar: to find you another position within the company. It is too early to discuss any other outcome as yet.*

Correct. This statement presents information that fits the client's primary need and provides information about the focus of the session. Because it is on topic and timely, the information is likely to be accepted by the client. Go to 10.9.

10.9

Effective information giving involves the ongoing use of basic attending skills: appropriate eye contact, body language, and verbal behavior. Good focusing and attending enhances the interviewer's awareness of the client's need for and reaction to information, facilitates the client's expression of emotions and attitudes, and maximizes the client's ability to integrate and use the information.

CLIENT: I appreciate that you are here to help me, but I'm not sure how all of this is going to turn out or even what these sessions are about, and I do have a family to support.

Choose the most appropriate response.

INTERVIEWER: You have a number of concerns right now. You're not sure about these sessions and how they will help with your job and your financial and family obligations. (Go to 10.10)

INTERVIEWER:	I can't help you with your other concerns; that's not my job. I'm here to help you identify an alternate job. Now, let's get on with it. (Go to 10.11)
INTERVIEWER:	It must give you a sense of relief that things are under control now. Perhaps we should get on, and try to find you another job. (Go to 10.12)

10.10 Your answer: *You have a number of concerns right now. You're not sure about these sessions and how they will help with your job and your financial and family obligations.*

Correct. You have identified and responded to the concerns expressed by the client. The use of basic attending skills increases the ability of the client to attend to information and facilitates the discussion of issues as they arise. Go to 10.13.

10.11 Your answer: *I can't help you with your other concerns; that's not my job. I'm here to help you identify an alternate job. Now, let's get on with it.*

This response provides information to the client but fails to respond to the concerns and needs expressed. The client's ability to use and integrate information effectively is dependent on the interviewer's ability to provide support for the client through the utilization of basic attending skills. Return to 10.9 and try again.

10.12 Your answer: *It must give you a sense of relief that things are under control now. Perhaps we should get on, and try to find you another job.*

Your attention to your own concern, getting to the task at hand, has resulted in your failure to attend to the client's emotional reaction. If the client's concerns and emotions are not recognized, they may continue to distract the client. As a result, subsequent information may not be integrated or utilized by the client. Return to 10.9 and try again.

10.13

Appropriate eye contact, voice tone, and body posture are important when giving information. In addition, clients are most likely to be receptive to information that is direct, clear, specific, concise, and concrete.

CLIENT:	Yes, I am concerned. I've been off work several months now, and I'd like to get back to work. I'm excited about these sessions. What are we going to do in them?

Choose the most appropriate response.

INTERVIEWER:	At this stage, I can't tell you much about what's involved. We have a lot of things to do in these sessions, and we really should get started. Maybe after we talk a little we can find you another position so that you can get back to work. (Go to 10.14)
INTERVIEWER:	What we can do right now is talk about some jobs that are available and that you might like to do. There are many jobs in the company—line workers, cleaners, clerks—a whole range of jobs. Maybe you would like one of these positions. What do you think? (Go to 10.15)
INTERVIEWER:	The purpose of these sessions is to find you an alternative position within the company. In order to do this, we need to identify your abilities, interests, and limitations. This will involve discussion and perhaps some testing. (Go to 10.16)

10.14 Your answer: *At this stage, I can't tell you much about what's involved. We have a lot of things to do in these sessions, and we really should get started. Maybe after we talk a little we can find you another position so that you can get back to work.*

When you give information, it is important that your response is direct, clear, specific, and concrete. Although this response is on topic, it is vague and indefinite with respect to what can be expected by the client. As such, it will do little to orient the client to what is to follow, and may even increase the client's apprehension of the process. Return to 10.13 and try again.

10.15 Your answer: *What we can do right now is talk about some jobs that are available and that you might like to do. There are many jobs in the company—line workers, cleaners, clerks—a whole range of jobs. Maybe you would like one of these positions. What do you think?*

This response provides the client with the type of information that an employee in a company knows already. The client has requested specific information. Your response is to provide concise, clear, and concrete information about what is to occur in the sessions. Return to 10.13 and try again.

10.16 Your answer: *The purpose of these sessions is to find you an alternative position within the company. In order to do this, we need to identify your abilities, interests, and limitations. This will involve discussion and perhaps some testing.*

Correct. This response is direct, clear, specific, concise, and concrete. As such, it will assist the client to integrate the information more easily and will contribute to the client's involvement in the sessions. Go to 10.17.

10.17 *When a client requests information, it is important that the interviewer match the information given to the client's ability to benefit from it. Good interviewers divide information into organized units rather than overload clients with information they are unable to integrate or use.*

CLIENT: I can understand why you would want to know about my abilities and limitations, but why do you want to know about my interests?

Choose the most appropriate response.

INTERVIEWER: Employees who enjoy and are satisfied by their work are more likely to remain content with their jobs. Your interests are important in identifying the type of work that will be satisfying to you. (Go to 10.18)

INTERVIEWER: Most individuals are unaware of their interests. In recent years, a number of excellent tests have been developed to help individuals identify their interests. The one we use here on a regular basis matches the interests of individuals like you to those of groups of people who are happy and contented with their jobs. It involves about 300 forced-choice questions and will take approximately 1 hour to complete. The test is sent away to be scored, and we will discuss the results when they are returned. (Go to 10.19)

INTERVIEWER: Interests have become very important in determining whether employees will stay with this company. We have a computerized assessment procedure that can give us detailed interest patterns of all employees and predictive information about their likely future with us. Currently we are working to validate these data. (Go to 10.20)

10.18 Your answer: *Employees who enjoy and are satisfied by their work are more likely to remain content with their jobs. Your interests are important in identifying the type of work that will be satisfying to you.*

Correct. You have provided the client with specific information in an amount that is appropriate to the request. The client will seek further information as required. Go to 10.21.

10.19 Your answer: *Most individuals are unaware of their interests. In recent years, a number of excellent tests have been developed to help individuals identify their interests. The one we use here on a regular basis matches the interests of individuals like you to those of groups of people who are happy and contented with their jobs. It involves about 300 forced-choice questions and will take approximately 1 hour to complete. The test is sent away to be scored, and we will discuss the results when they are returned.*

This information is too elaborate for the client's request. The information on testing is important, but is better given at a time just prior to the actual testing so that it will be useful to the client. Return to 10.17 and choose the response that contains only the information required by the client at this time.

10.20 Your answer: *Interests have become very important in determining whether employees will stay with this company. We have a computerized assessment procedure that can give us detailed interest patterns of all employees and predictive information about their likely future with us. Currently we are working to validate these data.*

At this stage in the interview, this information is far too extensive and complex for the client to absorb. It will simply overload and confuse the client and may lead to discomfort and a lack of cooperation. Return to 10.17 and try again.

10.21

To maximize the use of information by the client, it is important that the interviewer follow an information-giving response with a brief statement that checks out the degree to which the client has attended to the information provided. The interviewer can achieve this by asking the client to repeat the information, to act upon the information, or to engage in a discussion of the information. The client's response will enable the interviewer to judge the client's attention to the information given.

CLIENT: I've never thought about how important my interests were to my job—it does make sense, though.

Choose the most appropriate response.

INTERVIEWER: Let's move on now. I think we've discussed what you are here for long enough. I'm sure you're aware by now of the purpose of these sessions. (Go to 10.22)

INTERVIEWER: Before we proceed further, let me verify your understanding of what we are going to do. Could you tell me what you know about the purpose of these sessions? (Go to 10.23)

INTERVIEWER: We could spend a lot more time talking about the purpose of these sessions. I'm not sure whether we should do that. What do you think? (Go to 10.24)

10.22 Your answer: *Let's move on now. I think we've discussed what you are here for long enough. I'm sure you're aware by now of the purpose of these sessions.*

It is important that an interviewer check out the degree of client knowledge rather than assume it. Failure to check on a client's level of understanding can lead to incorrect assumptions and wasted time. Return to 10.21 and try again.

10.23 Your answer: *Before we proceed further, let me verify your understanding of what we are going to do. Could you tell me what you know about the purpose of these sessions?*

Correct. This response requests the client to summarize the information provided. It will assist the interviewer to determine the client's understanding of the information. The interviewer can then judge whether to provide more information or proceed with the interview. Go to 10.25.

10.24 Your answer: *We could spend a lot more time talking about the purpose of these sessions. I'm not sure whether we should do that. What do you think?*

This response checks on the client's wish for further discussion about the purpose of the interviews, but fails to check on whether the client has understood the information presented. Return to 10.21 and choose the response that checks on whether the client is attentive to the information discussed so far.

10.25   *It must be evident by now that information giving is not a simple task. The information given must match the client's needs and purposes. At any point in time, the client may request information that will help in the exploration, the understanding, or some elaboration of a topic.*

CLIENT: When I came in, I thought you were going to help me with my finances, but from what you've said, the first aim is to find me another position in the company. I understand the importance of assessing my abilities and interests, but what will happen if there isn't a job that matches them?

Choose the most appropriate response.

INTERVIEWER: I'm not sure we need to consider that now because it may not arise. Perhaps we should get on with the assessment process, and then if it does come up, we can talk about it. (Go to 10.26)

INTERVIEWER: If that occurs, you may have to decide between a position that, although not perfect, is close to your interests and abilities and a position located in another branch of the company. (Go to 10.27)

INTERVIEWER: You don't need to worry about that. This company has so many positions that there has to be one for you. There are vacancies that come up regularly in all departments. (Go to 10.28)

10.26 Your answer: *I'm not sure we need to consider that now because it may not arise. Perhaps we should get on with the assessment process, and then if it does come up, we can talk about it.*

This response denies the client's need to explore and further understand the situation. The information given should match the client's need to explore, understand, or seek elaboration of data or facts. Return to 10.25 and try again.

10.27 Your answer: *If that occurs, you may have to decide between a position that, although not perfect, is close to your interests and abilities and a position located in another branch of the company.*

Correct. Throughout the interviewing process, clients seek information that is for exploratory, understanding, or elaborative purposes. This response provides information that matches the client's need to explore a potential outcome and to understand further the situation within the company. Go to 10.29.

10.28 Your answer: *You don't need to worry about that. This company has so many positions that there has to be one for you. There are vacancies that come up regularly in all departments.*

The information contained in this response is false reassurance and does not meet the client's need to explore and understand what will occur if there is no match between the client's abilities and interests and the positions available. Return to 10.25 and try again.

10.29

It is not always possible for the interviewer to provide information in a form that matches the client's need. Sometimes clients have a particular bias or set arising from personal, emotional, or cultural experiences. Biases of this kind can result in the client's confusing, distorting, or rejecting the information given. Interviewers need to examine client statements for particular perspectives that reflect confusion, distortion, or rejection of information; through the use of good interviewing skills, interviewers can assist the client in developing a more realistic perspective.

CLIENT: So, it's most likely that you are going to make me move. If that happens, my family will have to give up their friends and activities, and we'd have to sell the house. I just cannot move; it would disrupt everything.

Choose the most appropriate response.

INTERVIEWER: You're really concerned about the long-term outcome of your situation and sensitive to the disruptions that might occur in your life. We need to review what we have talked about so far and try to put your concerns in perspective. (Go to 10.30)

INTERVIEWER: I didn't say you'd have to move; you've got it all wrong. I gave you moving as one option. It's not something that the company will require, it's just a possibility if you're interested. (Go to 10.31)

INTERVIEWER: If you have to move, the company has an excellent plan to assist you with the sale of your home and the purchase of a new one. There is even an employee-assistance program so that family members who have trouble adjusting can get counseling. (Go to 10.32)

10.30 Your answer: *You're really concerned about the long-term outcome of your situation and sensitive to the disruptions that might occur in your life. We need to review what we have talked about so far and try to put your concerns in perspective.*

Correct. This response identifies and reflects the emotional concern that is resulting in a distortion of the information presented. It also orients the client to a reality-based discussion directed toward resolution of the distortion. Go to 10.33.

10.31 Your answer: *I didn't say you'd have to move; you've got it all wrong. I gave you moving as one option. It's not something that the company will require; it's just a possibility if you're interested.*

This is a defensive response that could lead to an emotional interchange between client and interviewer. It does not assist the client to confront his distortion nor further the purpose of the interview. Return to 10.29 and try again.

10.32 Your answer: *If you have to move, the company has an excellent plan to assist you with the sale of your home and the purchase of a new one. There is even an employee-assistance program so that family members who have trouble adjusting can get counseling.*

This is an information-giving response, but the information will intensify further the client's distorted perspective. What is required is a response that enables the client to confront the distortion and make effective use of the information given. Return to 10.29 and try again.

10.33

One use of information is to orient the client to the interviewing process. Orienting statements are used to give information about the interviewer's and client's roles, and the costs, benefits, and experiences associated with the interviewing process. When orienting statements are used, clients are more likely to be receptive, to be motivated to continue, and to accomplish the tasks of the interview.

When orienting the client to the interview or an aspect of it, the interviewer presents the information in a detailed and concrete manner and with sufficient discussion to ensure that the information is integrated.

After resolving the client's distorted perspective on the availability and location of work within the company, the interviewer and the client have agreed that interest testing may provide the client with useful information.

Choose the most appropriate response.

INTERVIEWER: Interest testing is very important. It provides valuable information that will help to locate a new job for you. It's hard to describe what's involved, but as you complete the test and get the results, it will become clear. (Go to 10.34)

INTERVIEWER: Few people give any thought to their interests and how these relate to work. They are concerned only with how much money they will earn. People like this can end up very dissatisfied with their jobs. We don't want that to happen to you, so we'd better identify your interests before we go any further. (Go to 10.35)

INTERVIEWER: Interest testing is important in order to find out the alternative job-related interests you have. Interest testing provides information about basic interest patterns and identifies occupational groups that have interests similar to yours. (Go to 10.36)

10.34 Your answer: *Interest testing is very important. It provides valuable information that will help to locate a new job for you. It's hard to describe what's involved, but as you complete the test and get the results, it will become clear.*

This detailed orienting statement does not provide the client with any concrete information about interest testing. As a result, the client will be unclear about the reason for testing and may not benefit from the activity. Return to 10.33 and try again.

10.35 Your answer: *Few people give any thought to their interests and how these relate to work. They are concerned only with how much money they will earn. People like this can end up very dissatisfied with their jobs. We don't want that to happen to you, so we'd better identify your interests before we go any further.*

This response reflects the interviewer's belief about interests and work satisfaction, but does not provide the client with concrete information about interest testing. Return to 10.33 and choose the response that orients the client to the purpose of interest testing.

10.36 Your answer: *Interest testing is important in order to find out the alternative job-related interests you have. Interest testing provides information about basic interest patterns and identifies occupational groups that have interests similar to yours.*

Correct. This orienting statement presents information about interest testing that is brief, detailed, and concrete. It will result either immediately, or subsequent to further discussion, in the client's being aware of the importance of such testing in job relocation. Go to 10.37.

10.37   *Another form of informational response, the instruction or direction, is designed to assist the client to respond in a defined way. In an instruction, the interviewer provides information that is necessary for the client to acquire, strengthen, weaken, or eliminate a response. Instructions may include directions about test procedures, specific interventions, or methods of changing behavior. The interviewer presents this information in the form of a request, providing the client with directions about what to do, how to do it, and the potential outcome of appropriate and inappropriate performance.*

CLIENT: I never thought about it before, but I can see the importance of knowing about my interests. What do I have to do?

Choose the most appropriate response.

INTERVIEWER: Sit over at the table. Everything is set out for you. Read the instructions and when you understand what to do, begin the test. Let me know when you have finished. (Go to 10.38)

INTERVIEWER: I'd like you to sit over here. Please read the instructions carefully and check with me if you are uncertain about what to do. It is important to follow the instructions and complete the answer sheet carefully so that the information obtained will be accurate. If you answer carelessly, the test may not be scoreable or useful. Please tell me when you are finished. (Go to 10.39)

INTERVIEWER: Here is the test. Please complete it and hand it to the secretary when you're finished. (Go to 10.40)

10.38 Your answer: *Sit over at the table. Everything is set out for you. Read the instructions and when you understand what to do, begin the test. Let me know when you have finished.*
This instruction is in the form of a command that does not permit the client to clarify the instructions. Further, it does not acquaint the client with the potential benefits of compliance compared with the negative impact of noncompliance. Return to 10.37 and select the instruction that is more appropriate.

10.39 Your answer: *I'd like you to sit over here. Please read the instructions carefully and check with me if you are uncertain about what to do. It is important to follow the instructions and complete the answer sheet carefully so that the information obtained will be accurate. If you answer carelessly, the test may not be scoreable or useful. Please tell me when you are finished.*
Correct. This instruction requests the client to carry out the required task and provides information about what to do and how to do it, along with the consequences of carrying out the task appropriately and inappropriately. Go to 10.41.

10.40 Your answer: *Here is the test. Please complete it and hand it to the secretary when you're finished.* This abrupt response fails to provide the client with the information required to complete the task appropriately. When instructions are inadequate, the potential for unsatisfactory performance increases. Return to 10.37 and try again.

10.41   *Feedback is a form of information giving that confirms, changes, or widens a perspective held by the client. Effective feedback is concrete and specific, and whenever possible focuses upon the positive rather than the negative potential of the client. Sometimes interviewers are required to provide unpleasant information to a client. Whether the feedback is positive or negative, the interviewer presents the information as objective fact, delivered with the appropriate degree of supportive nonverbal behavior.*
By the next session, the results of the interest test have been returned. They reveal that the client has highly developed interest patterns in the field of mechanical activities, the area of employment no longer possible because of the work injury. The results indicate a secondary but less developed interest in the business-management area.

CLIENT: I'm really anxious to get the results of the interest test back. What are the results like?

Choose the most appropriate response.

INTERVIEWER: Your results are back, and here is your copy. Unfortunately, the results show that your strongest interests are in the area of mechanical activities. You have a secondary interest pattern in the business-management area. You could move into

	one of the positions in the purchasing department, but this would be difficult because of your education. What are your reactions to this? (Go to 10.42)
INTERVIEWER:	Your results are back, and here is your copy. Like you, I'm really excited about the results; they show a whole new area of interest. Have you ever thought of being a purchasing agent? The results show that you would be happy in this area after a little training. What are your reactions to this? (Go to 10.43)
INTERVIEWER:	Your results are back, and here is your copy. As you might expect, the results reveal that you have very strong interest patterns in the area of mechanical activities. More important, the results show that you also have interests in the business-management area and an occupational interest pattern similar to those of purchasing agents. What are your reactions to this? (Go to 10.44)

10.42 Your answer: *Your results are back, and here is your copy. Unfortunately, the results show that your strongest interests are in the area of mechanical activities. You have a secondary interest pattern in the business-management area. You could move into one of the positions in the purchasing department, but this would be difficult because of your education. What are your reactions to this?*

This response provides feedback to the client, but the emphasis is on the negative component of the information. Good interviewers present complete information, but place the emphasis on the positive potential of the client whenever possible. Return to 10.41 and try again.

10.43 Your answer: *Your results are back, and here is your copy. Like you, I'm really excited about the results; they show a whole new area of interest. Have you ever thought of being a purchasing agent? The results show that you would be happy in this area after a little training. What are your reactions to this?*

Feedback information is best delivered as objective fact. The overpositive enthusiasm of the interviewer can mislead the client into believing that the current occupational problem is resolved and that a position as a purchasing agent will be available in the company. Should this not be the case, this response will lead to further disappointment and difficulty for the client. Return to 10.41 and choose the response that is supportive but more objective.

10.44 Your answer: *Your results are back, and here is your copy. As you might expect, the results reveal that you have very strong interest patterns in the area of mechanical activities. More important, the results show that you also have interests in the business-management area and an occupational interest pattern similar to those of purchasing agents. What are your reactions to this?*

Correct. This feedback is concrete, specific, and objective. It focuses on the positive potential of the client while confirming the unpleasant fact that primary interests remain in the work area that is no longer possible. Go to 10.45.

10.45   *Another form of information giving is reframing. When reframing, the interviewer provides an alternative perspective that modifies, restructures, or gives new meaning to the client's perception of a situation or behavior. Reframes* *reduce defensiveness and enable clients to confront issues and begin to act in new ways. In forming the reframe, the interviewer considers what is acceptable to the client and then gives a new label or meaning to a situation or behavior.*

The interviewer and client have discussed the interest test results, have established that the client is willing to consider becoming a purchasing agent, and have established that a position in this area is available when retraining is completed. The client is hesitant to pursue this plan because of retraining.

CLIENT: I'm not sure about this retraining; I really don't want to return to school. I've been very happy doing what I've been doing. Why can't the company find me something where I don't have to retrain? I can't see going back to school. I can't see myself answering to a teacher. I don't like being told what to do.

Choose the most appropriate response.

INTERVIEWER: It's good to hear you talk about how happy you were with your past job, and how you valued and still value your independence. The purchasing agent position involves considerable independent thinking and action, and the training program is designed to facilitate the development of this independence. Your desire for independence and the goals of the training program appear to be complementary. (Go to 10.46)

INTERVIEWER: Given your previous job level in the company, the only way you can find a position that pays the same is if you retrain for another comparable position. You should be aware that the company will support you during your retraining at your previous salary. You must appreciate the benefit of this. (Go to 10.47)

INTERVIEWER: It's good to hear you were happy in your past job, but are you sure that you are not being stubborn in order to avoid the anxiety of evaluation at school? Often people avoid changes because they cannot predict what lies ahead. It seems that your hesitancy could be viewed as anxiety about change. (Go to 10.48)

10.46 Your answer: *It's good to hear you talk about how happy you were with your past job, and how you valued and still value your independence. The purchasing agent position involves considerable independent thinking and action, and the training program is designed to facilitate the development of this independence. Your desire for independence and the goals of the training program appear to be complementary.*

Correct. This response provides information based upon the client's comments that will enable the client to view retraining from a different, more adaptive perspective. Go to 10.49.

10.47 Your answer: *Given your previous job level in the company, the only way you can find a position that pays the same is if you retrain for another comparable position. You should be aware that the company will support you during your retraining at your previous salary. You must appreciate the benefit of this.*

This is important information for the client concerning remuneration while retraining, but it is premature. At this point, it is necessary to assist with the hesitancy, apprehension, and frustration of the client at having to retrain. A reframe that helps the client relabel these feelings in order to be able to engage in retraining is more facilitative. Return to 10.45 and try again.

10.48 Your answer: *It's good to hear you were happy in your past job, but are you sure that you are not being stubborn in order to avoid the anxiety of evaluation at school? Often people avoid changes because they cannot predict what lies ahead. It seems that your hesitancy could be viewed as anxiety about change.*

This response presents an alternative perspective to the client, but not necessarily one that will be acceptable to him without considerable discussion of the challenge you have presented. The essence of a reframe is to present the client with an alternative perspective that enables him to restructure his thinking and begin to act in a new way. Return to 10.45 and try again.

10.49 *Reframes can be positive but also deeply challenging. A new way of viewing the world can sometimes be painful as well as revealing and helpful. The test of the success of a reframe is whether the client thinks, feels, or behaves in a new or different way.*

CLIENT: It is true I like to be independent, and I can see myself participating in the training if it leads to a new and independent position. However, I still see the time it will take as being wasted time. I want to get going with the new job.

Choose the most appropriate response.

INTERVIEWER: I'm not sure the company would permit you to become a purchasing agent without the appropriate training. It is a highly specialized job and I can just imagine the reaction of the other agents if you were parachuted into their department without having the appropriate skills. You need to think seriously about committing to retraining before you take up the new position. (Go to 10.50)

INTERVIEWER: I am glad you are looking forward to an independent position. One way to look at retraining is that it can lead to a full and different life with new opportunities. New skills learned are never wasted. However, it is important to be patient and commit fully to the training program if you are going to succeed. If you are halfhearted about your commitment, it will be a waste of time for both you and the company. (Go to 10.51)

INTERVIEWER: It's good to hear that you're looking forward to getting back to work after the accident. The retraining period, however, can serve more than one purpose. Although it will provide you with new skills, you might want to consider the fact that it will also extend your recovery period and let you regain your former strength—a bonus for you—before taking on the responsibilities of a new and demanding job. (Go to 10.52)

10.50 Your answer: *I'm not sure the company would permit you to become a purchasing agent without the appropriate training. It is a highly specialized job and I can just imagine the reaction of the other agents if you were parachuted into their department without having the appropriate skills. You need to think seriously about committing to retraining before you take up the new position.*

It is unlikely that this response will convince the client that retraining is essential before he can commence his new position. A reframe should prompt him to consider a different way of viewing the retraining period, and change his response to it. Return to 10.49 and try again.

10.51 Your answer: *I am glad you are looking forward to an independent position. One way to look at re-training is that it can lead to a full and different life with new opportunities. New skills learned are never wasted. However, it is important to be patient and commit fully to the training program if you are going to succeed. If you are halfhearted about your commitment it will be a waste of time for both you and the company.*

This response begins with a reframe, but becomes a series of advice-laden statements that will not change the client's perspective about retraining. Return to 10.49 and try again.

10.52 Your answer: *It's good to hear that you're looking forward to getting back to work after the accident. The retraining period, however, can serve more than one purpose. Although it will provide you with new skills, you might want to consider the fact that it will also extend your recovery period and let you regain your former strength—a bonus for you—before taking on the responsibilities of a new and demanding job.*

Correct. This reframe reveals to the client the dual benefits of the retraining period, challenging him to think more deeply about that period. It is likely to assist the client to think differently about the benefits of the training that is being offered him. Go to 10.53.

10.53   *The most general form of information giving is the informational response. Informational responses are used frequently after clients have determined their goals. During this phase of the interviewing process, clients require information to broaden the alternatives generated, to identify potential outcomes of particular alternatives, and to correct any invalid or unreliable information held. Informational statements may give the client direct information, direct the client to other sources of information, or refer the client to people with relevant experience. Clients understand and appreciate information better if they are involved in its acquisition. Good interviewers frame informational statements in a way that summarizes the client's requirements and the means by which these can be met.*

After a discussion of the independence involved in the purchasing agent job, the client has decided that retraining leading to a position in purchasing is acceptable. The client is now in a position to seek more information about what is required before job relocation occurs.

CLIENT: I hadn't thought of the added benefits of the retraining period. It looks as if retraining will not only teach me a new skill but also will give me more time to recover from my injury. I think I need that time. Maybe you could tell me exactly what's involved in retraining, and what I will be doing when I'm finished.

Choose the most appropriate response.

INTERVIEWER: Now that you are more comfortable with a move to a new position, you require information about what to expect. I'm not sure what's involved. Probably, you will be sent out to college and have on-the-job training as well, or maybe all you will need is on-the-job training. There are many activities that go on in purchasing that you'll need to become familiar with. The company has arrangements with a number of suppliers; certainly, you'll need to learn all about those. (Go to 10.54)

INTERVIEWER: Now that you are more comfortable with a move to a different position, you require information about what to expect. Possibly, the most effective way to achieve this is to have you meet with the manager of the purchasing department to discuss both retraining and activities involved in positions there. What do you think about this? (Go to 10.55)

INTERVIEWER: I don't know much about what's involved in purchasing. If you're interested, maybe you should call the manager of that department and set up an interview. I think we have completed all that we can do. Give me a call if you think I can help further. (Go to 10.56)

10.54 Your answer: *Now that you are more comfortable with a move to a new position, you require information about what to expect. I'm not sure what's involved. Probably, you will be sent out to college and have on-the-job training as well, or maybe all you will need is on-the-job training. There are many activities that go on in purchasing that you'll need to become familiar with. The company has arrangements with a number of suppliers; certainly, you'll need to learn all about those.*

The client is requesting direct information that will provide a sense of security about the decision made. Good interviewers direct clients to resources that can provide accurate and valid information. This response fails to provide specific information. It raises more questions than it answers. Return to 10.53 and try again.

10.55 Your answer: *Now that you are more comfortable with a move to a different position, you require information about what to expect. Possibly, the most effective way to achieve this is to have you meet with the manager of the purchasing department to discuss both retraining and activities involved in positions there. What do you think about this?*

Correct. This response summarizes the client's concern, identifies an appropriate resource, and solicits the client's reaction to what is suggested. Clients benefit most from information when they are directly involved in its acquisition. Go to 10.57.

10.56 Your answer: *I don't know much about what's involved in purchasing. If you're interested, maybe you should call the manager of that department and set an interview. I think we have completed all that we can do. Give me a call if you think I can help further.*

This response fails to summarize the client's concern. Although a resource is suggested, the manner of presentation is unlikely to facilitate the client's use of the resource. The abrupt termination of the interaction leaves the client with insufficient information and may hamper progress toward the goal of job relocation. Return to 10.53 and try again.

10.57   Review Questions

Indicate whether each of the following statements is true or false.

1. The amount of information provided should be matched to the client's ability to benefit from it.

2. The interviewer is powerless to correct a client's distortion of the information provided.
3. Other interviewing responses have no role during information giving and integration.
4. Orienting responses have the effect of involving clients in the interviewing process.
5. Information should be provided when the client is most receptive to it.
6. Reframing involves the presentation of an alternative perspective from which the client can view a situation.
7. Instructions should never include information about potential outcomes.
8. Information-giving responses are based upon the interviewer's personal experience and opinion.
9. The client's attention to the information given can be evaluated by requesting the client to repeat, act upon, or discuss it.
10. When presenting information, the interviewer should provide only broad and general facts.

Review Answers

1. *True.* If you answered *false,* go to 10.17 and review.
2. *False.* If you answered *true,* go to 10.29 and review.
3. *False.* If you answered *true,* go to 10.9 and review.
4. *True.* If you answered *false,* go to 10.33 and review.
5. *True.* If you answered *false,* go to 10.5 and review.
6. *True.* If you answered *false,* go to 10.45 and review.
7. *False.* If you answered *true,* go to 10.37 and review.
8. *False.* If you answered *true,* go to 10.1 and review.
9. *True.* If you answered *false,* go to 10.21 and review.
10. *False.* If you answered *true,* go to 10.13 and review.

If six or more of your responses were incorrect, you should return to 10.1 and review the material in this chapter.

Points to Remember about Information Giving

1. WHEN GIVING INFORMATION:
   a. Provide data or facts relevant to the client's need.
   b. Ensure that the client is receptive to the information.
   c. Be direct, clear, specific, concise, and concrete.
   d. Break the information into units that the client can utilize.
2. AFTER GIVING INFORMATION:
   a. Check that the client has attended to the data and facts provided.
   b. Evaluate for distortions and use other interviewing skills to correct them.

3. USE INFORMATION GIVING:
   a. To orient clients to the interviewing process.
   b. To provide instructions or directions.
   c. To present feedback.
   d. To provide alternative perspectives.
   e. To direct clients to other resources.

10.1

Activity Unit

Find two others who have also read this chapter and role play short interviews alternating the client, interviewer, and observer roles. Record each brief interview on audio- or videotape. An interview with at least five interviewer responses is sufficient for this exercise.

Interviewer role Practice information giving in response to the client.

Client role Be cooperative and seek information from the interviewer. Clients may want to role play the problem of a friend or family member that requires information for its solution or to enact one of the following roles:

1. A parent whose child has been diagnosed as chronically ill.
2. A person whose home has just been lost in a flood.
3. An employee discussing the possibilities of a job change within an organization.
4. A person who finds it hard to organize his or her work and feels overwhelmed.

Observer role Operate the equipment and complete the Practice Interview Checklist (see Table 10.2).

Postinterview discussion Using the taped interview and Practice Interview Checklist data, the participants should discuss the performance of the interviewer. All participants can learn from this discussion, which should be nonjudgmental, focus on the positive as well as less effective responses, and stimulate improved performance for all.

Table 10.2

Practice Interview Checklist

Activity	Interviewer response[a]				
	1	2	3	4	5
Appropriate nonverbal behavior					
On topic					
Orienting statement					
Instruction or direction					
Feedback					
Alternate perspective (reframe)					
Informational statement					
Effective response[b]					

[a] Check the activities that characterize each response.

[b] Indicate the degree of effectiveness on a scale from 1 to 5 in which 1 is ineffective and 5 is very effective.

10.2

Activity Unit

Using the tape of the interview produced in Activity Unit 10.1 or the tape of another practice interview, enter the original interview responses below. Then formulate a better response to replace each original one in the space provided.

Original response 1 _____

_____

_____

Better response _____

_____

_____

Original response 2 _____

_____

Better response _____

_____

_____

Original response 3 _____

_____

_____

Better response _____

_____

_____

Original response 4 _____

_____

_____

Better response _____

_____

_____

Original response 5 _____

_____

_____

Better response _____

_____

_____

Cultural Considerations

1. Ensure that the information given is culturally appropriate and relevant.
2. Many cultures view information as an important vehicle for change and expect the interviewer to be a source of information.
3. In cultures where it is important to save face, the use of positive reframing is important.

InfoTrac® College Edition Keyword Search Terms

Information Giving AND Communication
Information Giving AND Counseling
Feedback AND Communication
Feedback AND Counseling

Interpreting

*T*his chapter is intended to help you master the skill of interpreting. After completing this chapter, you should be able to:

1. Identify the underlying meaning of the client's narrative or story.
2. Formulate an interpretation that enables the client to view the narrative from an alternative perspective.
3. Decide when and how to deliver an interpretation.
4. Facilitate the client's effective use of an interpretation.

The interpreting response is different from other interviewing responses in that the interviewer focuses not on the explicit message of the client but on the underlying, implicit message the client is communicating. Although there has been some controversy about the use of interpretation as an interviewing skill, more recent information-processing and multicultural theories argue for its value in enhancing the client–interviewer relationship, in assisting clients to perceive things differently, and in facilitating change in client behavior. More specifically, interpretation has been found to promote client disclosure, enhance interviewer credibility, and increase client self-understanding and goal setting. Research findings support the appropriate use of interpreting as an effective interviewing response.

Each frame in this chapter is complete in itself; that is, each focuses on a different client in a different situation. To give you experience with a range of interpretations, some frames may have more than one correct response. Go to 11.1.

11.1

An interpretation is the expression of a hypothesis about the issue confronting the client. It is formulated by the interviewer on the basis of the client's story. It is the active interviewing skill of communicating to a client an alternative meaning inherent in the client's communications to the interviewer. Interpreting is an intervention that goes beyond the explicit story the client is communicating, encouraging the client to consider an alternative explanation for events, behaviors, feelings, attitudes, or thoughts. There are two steps in interpreting: (1) identifying the client's implicit message, and (2) presenting to the client this implicit message as a different way of viewing the situation.

To make an interpreting response, it is important to first identify the implicit message in the client's story. The next series of frames helps you develop the ability to identify the implicit message the client is communicating.

CLIENT: (a 50-year-old woman discussing her current health concerns with her family physician) What really brought me to see you are my problems in sleeping. I lie awake at night trying to remember the things that I'm supposed to be doing the next day. I have become very forgetful. I almost created a fire the other day when I forgot to turn off the electric stove. I don't know what's happening to me.

Choose the statement that best represents the implicit message in the client's story.

Concern about her inability to sleep (Go to 11.2)

Concern about a serious illness (Go to 11.3)

Concern about remembering things (Go to 11.4)

11.2 **Your answer:** *Concern about her inability to sleep*
You have attended to an explicit part of the client's story, but have failed to identify an implicit meaning. Return to 11.1 and try again.

11.3 **Your answer:** *Concern about a serious illness*
Correct. You have identified an implicit message in the client's story. Go to 11.5.

11.4 **Your answer:** *Concern about remembering things*
You have identified one of the messages in the client's story, but not the implicit message. Return to 11.1 and try again.

11.5	CLIENT:	(a single father with two school-age children) I work hard all day, and when I come home, the children are waiting for me. I love them very much, but they are very demanding. They never stop asking me to do things. I try to get them to play together, but all they want is to play with me. I'm so frustrated. I feel so helpless.

Choose the statement that best represents the implicit message in the client's story.

Sorry for having any children (Go to 11.6)

Joy that his children love him (Go to 11.7)

Guilty at wanting time for himself (Go to 11.8)

11.6 Your answer: *Sorry for having any children*
There is no evidence for this implicit message in the client's story. Return to 11.5 and try again.

11.7 Your answer: *Joy that his children love him*
This may be true, but there is no evidence for this implicit message in the client's story. Return to 11.5 and try again.

11.8 Your answer: *Guilty at wanting time for himself*
Correct. You have identified an implicit message in the client's story. Go to 11.9.

11.9	CLIENT:	(the parent of two teenagers married to an alcoholic) I've tried to make this marriage work. I know it's useless, but I can't seem to let go. When I think about leaving, I get a real sense of panic. I guess I'll have to make the best of things.

Choose the statement that best represents the implicit message in the client's story.

Scared at what the future holds if separation occurs (Go to 11.10)

Exhausted by the effort of coping with an alcoholic partner (Go to 11.11)

Committed to the marriage (Go to 11.12)

11.10 Your answer: *Scared at what the future holds if separation occurs*
Correct. You have identified an implicit message in the client's story. Go to 11.13.

11.11 Your answer: *Exhausted by the effort of coping with an alcoholic partner*
This may indeed be the case, but there is no evidence of this in the client's story. Return to 11.9 and try again.

11.12 Your answer: *Committed to the marriage*
This is not evident in either the client's implicit or explicit message. Return to 11.9 and try again.

11.13

The second step in making an interpreting response is to present the implicit message inherent in the client's communication so that the client can consider the situation from a different viewpoint.

CLIENT: (a female employee meeting with a supervisor who wishes to discuss the employee's ongoing lack of productivity on the job) I'm trying to do my work, but I'm just so tired. It is very difficult to concentrate on my job with the hours I spend managing and coaching the rowing team. That's so rewarding.

Choose the most appropriate response.

INTERVIEWER: Given your difficulties at work, it might be advisable to think about giving up your rowing commitments. You are an important employee, but your work has deteriorated badly. (Go to 11.14)

INTERVIEWER: You seem to be saying that your rowing commitments have greater priority in your life than your work right now. Perhaps understanding this will help you identify and change your work behavior. (Go to 11.15)

INTERVIEWER: It must be satisfying to you as a coach to have such an interesting pastime, especially when the team keeps winning the way it is. As manager of the team, I am sure you appreciate the importance of each team member's contribution. (Go to 11.16)

11.14 Your answer: *Given your difficulties at work, it might be advisable to think about giving up your rowing commitments. You are an important employee, but your work has deteriorated badly.*

This response ignores the client's implicit message and fails to encourage the client to consider any alternative explanation for her behavior. Return to 11.13 and try again.

11.15 Your answer: *You seem to be saying that your rowing commitments have greater priority in your life than your work right now. Perhaps understanding this will help you identify and change your work behavior.*

Correct. This interpreting response identifies the implicit message in the client's statement—the priority of rowing—and provides her with another explanation for her attitude and behavior. The interviewer has verbalized an issue that the client may not have recognized. Go to 11.17.

11.16 Your answer: *It must be satisfying to you as a coach to have such an interesting pastime, especially when the team keeps winning the way it is. As manager of the team, I am sure you appreciate the importance of each team member's contribution.*

This response does not give any indication that you have identified the implicit message in what the client has communicated to you. Unless you can first identify the implicit message, you will be unable to develop an appropriate interpretation. Return to 11.13 and try again.

11.17

Interpreting responses are based on what the interviewer decides is causing or contributing to the client's difficulties or undesirable situation. Interviewers make interpretations when they assist clients to make connections between disparate statements and events, identify patterns of behavior and recurring themes in the discussion, relate present occurrences to past events, and develop a new orientation to their feelings and behaviors, including their impact on others. Interviewers use personal observations, common sense, and often their theoretical perspectives when interpreting.

CLIENT: (a high school student with a history of poor performance, who fails to complete assigned work, misses classes, and when in class, tends to be disruptive; the student is meeting with the school principal to discuss future schooling) My classes are really boring, my schedule is a mess, and I don't like the others in my classes. Having math first thing in the morning is stupid. I want to be at school, but I can't get here for early classes. Getting to school on time has always been a problem. There's no one to wake me. My mom goes to work early.

Choose the most appropriate response.

INTERVIEWER: I might be wrong, but I wonder if some of your difficulty lies in the fact that you have never learned how to use the educational system to your personal advantage. (Go to 11.18)

INTERVIEWER: You seem to be saying that your responsibilities at home interfere with your full involvement here at school. (Go to 11.19)

INTERVIEWER: As I think about what you are saying, it occurs to me that perhaps you haven't as yet decided to take responsibility for your own education. (Go to 11.20)

11.18 Your answer: *I might be wrong, but I wonder if some of your difficulty lies in the fact that you have never learned how to use the educational system to your personal advantage.*
Correct. You have identified one possible message implicit in what the client is saying based upon your perception of the meaning of past and present events in this student's life. You have provided a different meaning to the message portrayed by the student. If this is your first correct response, return to 11.17 and identify another appropriate response before going to 11.21.

11.19 Your answer: *You seem to be saying that your responsibilities at home interfere with your full involvement here at school.*
This is an interpretation, but your inference of the client's implicit message goes well beyond the information provided. As such, it could anger the student, resulting in defensiveness and lack of cooperation. Return to 11.17 and try again.

11.20 Your answer: *As I think about what you are saying, it occurs to me that perhaps you haven't as yet decided to take responsibility for your own education.*
Correct. This response focuses on an implicit message of the client pointing out the possible association between the stated wish to be at school and the lack of

understanding about the behavior that is necessary for this to occur. By focusing on this association, you give the client a new orientation to the problem. If this is your first correct response, return to 11.17 and identify another appropriate response before going to 11.21.

11.21

The purpose of an interpreting response is to provide clients with a broader understanding of their behaviors, attitudes, thoughts, and feelings. This broader understanding enables clients to make different choices, engage in new, more functional behaviors, and obtain an increased sense of responsibility and control.

CLIENT: (a male client who has been meeting with a nurse in an employee wellness program for several weeks in order to stop smoking, but who has failed to complete any of the planned strategies) I'm afraid I haven't had time to set up the monitoring program that you suggested. I thought about it, but I couldn't find the cards that you said I should use.

Choose the most appropriate response.

INTERVIEWER: My observations of your progress so far suggest that you may not be ready to commit to giving up smoking. You seem to be at a stage of contemplating whether or not you will make this change in your life. (Go to 11.22)

INTERVIEWER: I find it hard to believe that you couldn't find cards if you wanted to. There are so many stores that sell them. You really must make an effort if you wish to stop smoking. (Go to 11.23)

INTERVIEWER: It seems as if stopping smoking is important to you, but that you need more time than I'm giving you to complete the assignments. Perhaps we should meet every other week instead of weekly. (Go to 11.24)

11.22 Your answer: *My observations of your progress so far suggest that you may not be ready to commit to giving up smoking. You seem to be at a stage of contemplating whether or not you will make this change in your life.*

Correct. This response identifies the implicit message of the client—his lack of commitment to action around smoking reduction. By interpreting to the client his possible stage in the change process, the interviewer assists the client to develop a broader understanding of and increased responsibility for his behavior. Go to 11.25.

11.23 Your answer: *I find it hard to believe that you couldn't find cards if you wanted to. There are so many stores that sell them. You really must make an effort if you wish to stop smoking.*

This response focuses on only one aspect of the client's message—the cards. This has little relevance to what is preventing the client from progressing in the program. You need to identify the implicit message in what the client is communicating to formulate an interpretation that will assist the client to work toward making an important change in his life. Return to 11.21 and try again.

11.24 Your answer: *It seems as if stopping smoking is important to you, but that you need more time than I'm giving you to complete the assignments. Perhaps we should meet every other week instead of weekly.*

This response identifies the surface meaning of what the client is saying, but not the implicit message. This response and the proposed action have little potential of assisting the client to move toward his goal. Return to 11.21 and try again.

11.25

Following an interpretation, it is good practice for the interviewer to examine its effectiveness. A client may accept, fail to comprehend, or reject the interviewer's interpretation. The client's response may be expressed verbally or nonverbally. When there is no voluntary response by the client, the interviewer has the responsibility for checking the accuracy of the interpretation.

This sequence of responses continues from the previous frame.

CLIENT: (with a puzzled, anxious look, but offering no response)

Choose the most appropriate response.

INTERVIEWER: You seem uncomfortable with my last comment. Now let's talk about how to move forward from here. (Go to 11.26)

INTERVIEWER: You seem uncomfortable with my last comment. I'm only trying to assist you to stop smoking. (Go to 11.27)

INTERVIEWER: You seem uncomfortable with my last comment. Can you share your reactions to what I've just said? (Go to 11.28)

11.26 Your answer: *You seem uncomfortable with my last comment. Now let's talk about how to move forward from here.*

You have identified the client's nonverbal reaction to your interpretation. Rather than following up on the impact of the interpretation, you have chosen to move forward. This failure to pursue the interpretation and understand the client's reaction to it could negate its utility in the interviewing process. Return to 11.25 and try again.

11.27 Your answer: *You seem uncomfortable with my last comment. I'm only trying to assist you to stop smoking.*

You have identified the client's nonverbal response to the interpretation. However, you have failed to initiate further discussion of the interpretation. Further discussion may lead to its confirmation, modification, or rejection in favor of an alternative interpretation. Return to 11.25 and try again.

11.28 Your answer: *You seem uncomfortable with my last comment. Can you share your reactions to what I've just said?*

Correct. You have identified the client's discomfort with the response and have used an open question to seek information about the client's perceived accuracy of the interpretation. Go to 11.29.

11.29

When making an interpreting response, it is advisable to formulate and deliver the response in a tentative rather than an absolute manner. A tentative response facilitates client exploration and is less likely to result in a negative response from the client.

CLIENT: (a female client, an employee in a large corporation undergoing restructuring, who is meeting with an employee-assistance counselor to discuss her difficulties at work) I'm becoming really worried. My job is important to me. I come to work each day determined to complete all the work that is piling up, but when I look back at the end of the day, I seem to have done very little. I think there's something wrong with my ability to concentrate.

Choose the most appropriate response.

INTERVIEWER: Your difficulties are associated with the threat to your job as restructuring occurs. You are unable to concentrate because you are preoccupied with thoughts of what you will do if you are laid off. (Go to 11.30)

INTERVIEWER: One possible explanation you may want to consider is that your lack of concentration is associated with your concern that your job is threatened by the restructuring that is going on in the corporation. What do you think? (Go to 11.31)

INTERVIEWER: I don't think being scared about restructuring is going to help you keep up with your workload. It is important for you to resolve this quickly before the problem gets worse. (Go to 11.32)

11.30 Your answer: *Your difficulties are associated with the threat to your job as restructuring occurs. You are unable to concentrate because you are preoccupied with thoughts of what you will do if you are laid off.*
You have identified an implicit message in the client's communication, but your interpretation is too direct. Your interpretation is more likely to be accepted for discussion by the client if it is more tentative. Return to 11.29 and try again.

11.31 Your answer: *One possible explanation you may want to consider is that your lack of concentration is associated with your concern that your job is threatened by the restructuring that is going on in the corporation. What do you think?*
Correct. This interpretation is phrased tentatively and will enable the client to consider it in a meaningful way. Go to 11.33.

11.32 Your answer: *I don't think being scared about restructuring is going to help you keep up with your workload. It is important for you to resolve this quickly before the problem gets worse.*
Although you have identified an implicit message in the client's response—her concern about restructuring—your emphatic and directive manner of presentation is likely to result in a negative response from the client. Return to 11.29 and try again.

11.33

An interpreting response is most likely to be understandable and result in change when it is slightly different from the client's viewpoint. When an interpreting response is highly discrepant from the client's viewpoint, it is more likely to be considered unacceptable and rejected.

CLIENT: (a male client, who is a workaholic, is seeking assistance because he has difficulty maintaining friendships; he is close to his mother but distant from his father) Over the years, I've tried to develop a number of friendships, but no one seems to stay friends with me for very long.

Choose the most appropriate response.

INTERVIEWER: From what you've said, it seems that your strong need for security based on your rejection by your father is interfering with your efforts to develop friendships. (Go to 11.34)

INTERVIEWER: It seems to me that your strong attachment to your mother is interfering with your ability to develop lasting friendships. What do you think? (Go to 11.35)

INTERVIEWER: From what you've said, it seems that one of the things you might want to think about is the degree to which your work habits interfere with your efforts to establish friendships. (Go to 11.36)

11.34 Your answer: *From what you've said, it seems that your strong need for security based on your rejection by your father is interfering with your efforts to develop friendships.*
This interpreting response is distant from the message conveyed to this point by the client. When interpretations are too discrepant from the client's implicit message, they are likely to be rejected and/or provoke a negative reaction from the client. Return to 11.33 and try again.

11.35 Your answer: *It seems to me that your strong attachment to your mother is interfering with your ability to develop lasting friendships. What do you think?*
This response is too distant from the client's stated viewpoint of his difficulty. As a result, it will not be understandable to him and is likely to be rejected. Return to 11.33 and try again.

11.36 Your answer: *From what you've said, it seems that one of the things you might want to think about is the degree to which your work habits interfere with your efforts to establish friendships.*
Correct. This interpreting response is close to the client's awareness and only slightly discrepant from the client's frame of reference. As a result, it is likely to be understandable to the client and be of use to him. Go to 11.37.

11.37

When formulating an interpreting response, the interviewer ensures that the response is based on adequate information shared by the client, is framed positively, and is presented in a manner that enlists the client's cooperation. The effective interviewer refrains from projecting biases and values and from making interpretations that provide excuses for clients.

CLIENT: (a female college student who has requested assistance because she is dissatisfied with her academic progress) I was home on the weekend and became really down. All that my parents talked about was my brother who is in the space program and my sister who's a resident in neurology. They didn't seem to be interested when I tried to talk about what I wanted to do.

Choose the most appropriate response.

INTERVIEWER: The impression that I am getting is that you would like your parents to be involved with you and display the same feelings to you that they do toward your brother and sister. (Go to 11.38)

INTERVIEWER: Some brothers and sisters are born lucky. All of the breaks seem to go their way, no matter how hard others try. When there is this pattern of family behavior, it's best to distance yourself from it all. (Go to 11.39)

INTERVIEWER: It appears to you that your parents are not interested in you or anything you do. It is unlikely you will be able to change this pattern of behavior. (Go to 11.40)

11.38 Your answer: *The impression that I am getting is that you would like your parents to be involved with you and display the same feelings to you that they do toward your brother and sister.*

Correct. This interpreting response is based on adequate information, is presented positively and is likely to engage the client in meaningful discussion of the issue. Go to 11.41.

11.39 Your answer: *Some brothers and sisters are born lucky. All of the breaks seem to go their way, no matter how hard others try. When there is this pattern of family behavior, it's best to distance yourself from it all.*

This interpreting response provides an excuse for the client, which if adopted is unlikely to facilitate positive change. The concluding comments suggest a biased coping strategy on the part of the interviewer. Return to 11.37 and try again.

11.40 Your answer: *It appears to you that your parents are not interested in you or anything you do. It is unlikely you will be able to change this pattern of behavior.*

This interpreting response is framed negatively and includes information that reflects a biased viewpoint of the outcome. It is unlikely to assist the client or change her perspective or behavior. Return to 11.37 and try again.

11.41

An interpreting response is most likely to assist the client when it is offered in the context of a facilitative relationship and when the client has demonstrated a willingness to explore issues. The timing of an interpreting response within a session is important, as some interpretations result in a negative or emotional response. For the client to discuss the interpretation in a meaningful way, the interviewer must present the interpretation at a point when sufficient time is available for facilitative discussion. If presentation of the interpreting response is to occur late in the session, it is preferable to postpone it until the next session.

CLIENT: (an athlete who has begun to meet with a sport psychologist in order to improve performance) I thought through what we talked about last week and it really helped me. I still don't know, though, what happens when I'm winning and everything falls apart. It's happened all my life, even when I was a child. My father used to tell me that I'd never be a winner.

Choose the most appropriate response.

INTERVIEWER: (midway through a session) You appear to be concerned that the problem is genetic and you will be unable to solve it. (Go to 11.42)

INTERVIEWER: (midway through a session) You mention your father and his influence frequently, and increasingly I find myself wondering if you're living out his expectations instead of your own. I wonder if his influence has anything to do with your difficulty in winning. What do you think? (Go to 11.43)

INTERVIEWER: We must finish the session shortly, but before we do, I want you to react to the following. You mention your father and his influence frequently, and increasingly I find myself wondering if you're living out his expectations instead of your own. I wonder if his influence has anything to do with your difficulty in winning. (Go to 11.44)

11.42 Your answer: *(midway through a session) You appear to be concerned that the problem is genetic and you will be unable to solve it.*

This interpretation goes well beyond the information you have available at this point. It also has the potential to provide the client with an excuse for poor performance. Return to 11.41 and try again.

11.43 Your answer: *(midway through a session) You mention your father and his influence frequently, and increasingly I find myself wondering if you're living out his expectations instead of your own. I wonder if his influence has anything to do with your difficulty in winning. What do you think?*

Correct. This interpreting response is offered to a client whose remarks demonstrate a willingness and an ability to explore issues. The timing of the response, midway through the session, enables you to support the client in meaningful discussion of the interpretation. Go to 11.45.

11.44 Your answer: *We must finish the session shortly, but before we do, I want you to react to the following. You mention your father and his influence frequently, and increasingly I find myself wondering if you're living out his expectations instead of your own. I wonder if his influence has anything to do with your difficulty in winning.*

It is unwise to share an interpretation late in a session no matter how important it is. Interpretations provoke a variety of reactions in clients. Time must be left for the interpretation to be considered and discussed with the client. Return to 11.41 and try again.

11.45

Clients react to an interpreting response in a variety of ways. These include meaningful discussion, rejection, thoughtfulness, silence, and a range of emotions. When clients reject an interpreting response or react with negative emotion, the interviewer may find it necessary to drop the rejected interpreting response temporarily, review it for accuracy, and then await a later opportunity to reintroduce it. Often, an interpretation needs to be repeated many times and in different ways before a client is able to assimilate it.

This sequence of responses continues from the previous frame.

CLIENT: (in a hostile manner) I don't think my failure to win has anything to do with my father. I loved my father very much. He's dead now, but he was a good father to me.

Choose the most appropriate response.

INTERVIEWER: It seems to me you're very protective of your father and not willing to look at his influence on your performance. (Go to 11.46)

INTERVIEWER: You have a high regard for your father and you feel your failure to win has other causes. (Go to 11.47)

INTERVIEWER: Your anger suggests that your father and his expectations for you may indeed be relevant to the difficulties you are having. (Go to 11.48)

11.46 Your answer: *It seems to me you're very protective of your father and not willing to look at his influence on your performance.*

When a client reacts negatively to an interpretation, it is wise to deal with that reaction before pursuing further interpretation of the client's story. Even if it were appropriate at this point, this interpretation is somewhat direct and, hence, likely to provoke another negative response from the client. Return to 11.45 and try again.

11.47 Your answer: *You have a high regard for your father and you feel your failure to win has other causes.*

Correct. When a client acts with hostility or rejects an interpreting response, it is advisable to put aside the interpretation and, having reviewed it for accuracy, await another time for the theme to arise. Themes and repetitive patterns of behavior are often the subject of interpretations that need to be repeated in a variety of ways before acceptance by the client. Go to 11.49.

11.48 Your answer: *Your anger suggests that your father and his expectations for you may indeed be relevant to the difficulties you are having.*

When a client reacts negatively to an interpretation, it is wise to put it aside temporarily and respond to the client's reaction. Having considered the accuracy of the interpretation, you can reintroduce it at a later point in a slightly different form. Return to 11.45 and try again.

11.49

The skill of interpreting is complex and, unlike most other interviewing skills, comes from the interviewer's perspective rather than the client's. Two or three skillful interpretations per session are usually the maximum number a client can use successfully. Beginning interviewers with little experience should use the skill with caution, ensuring each time that they have developed an accurate interpretation. Before using the skill of interpreting with a client, the interviewer should consider the client's self-esteem, ability to understand and use interpretation, and whether or not the client presents with a degree of emotional disturbance that mitigates against its use. Not all clients are able to profit from interpretations. Go to 11.50.

11.50

Review Questions

Indicate whether each of the following statements is true or false.

1. An interpreting response results in change when it is distant from the client's viewpoint.
2. The purpose of an interpreting response is to reduce the client's responsibility and control.
3. An interpreting response is based on the client's implicit message, enabling the client to consider the situation from another perspective.
4. It is advisable to present an interpreting response in a tentative manner.
5. Interpreting is a simple skill that all beginning interviewers should use.
6. It is good practice for the interviewer to check on the accuracy of the interpretation.
7. The timing of an interpreting response within the interviewing process is of no relevance.
8. It is unimportant whether an interpreting response is framed positively or negatively.

Review Answers

1. *False.* If you answered *true,* go to 11.33 and review.
2. *False.* If you answered *true,* go to 11.21 and review.
3. *True.* If you answered *false,* go to 11.1 and review.
4. *True.* If you answered *false,* go to 11.29 and review.
5. *False.* If you answered *true,* go to 11.49 and review.
6. *True.* If you answered *false,* go to 11.25 and review.
7. *False.* If you answered *true,* go to 11.41 and review.
8. *False.* If you answered *true,* go to 11.37 and review.

If five or more of your responses were incorrect, you should return to 11.1 and review the material in this chapter.

Points to Remember about Interpreting

1. AN INTERPRETING RESPONSE:
   a. Is based on the interviewer's view of the client's story.
   b. Identifies for the client relationships among events, patterns of behavior, themes discussed, and the interviewer's personal observations of the client.
   c. Encourages the client to consider an alternative explanation for events, behaviors, feelings, attitudes, or thoughts.
   d. Facilitates the development of new alternatives, more functional behaviors, and increased responsibility.

2. WHEN INTERPRETING:
   a. Ensure that a facilitative relationship has been established.
   b. Identify the implicit message inherent in the client's story.
   c. Present the implicit message to the client for consideration.
   d. Check the accuracy of the interpretation and engage the client in meaningful discussion of the interpretation.
3. WHEN MAKING AN INTERPRETING RESPONSE:
   a. Deliver the response in a tentative manner.
   b. Formulate a response that is only slightly discrepant from the client's communication.
   c. Frame the response positively and avoid responses that provide excuses for clients.
   d. Refrain from projecting biases and values.
   e. Ensure there is adequate time in the session to discuss the interpretation.
   f. Be prepared for a negative or emotional response from the client.

11.1

Activity Unit

Working with two others who have also read this chapter, role play short interviews alternating the client, interviewer, and observer roles. Record each brief interview on audio- or videotape. An interview with five interviewer responses, including one or more interpretations, is sufficient for this exercise.

Interviewer role Practice interpreting in combination with other basic interviewing skills.

Client role Be cooperative and facilitate the progress of the interview so the interviewer can practice the skill of interpreting. Clients may want to role play the problem of a friend or family member or to enact one of the following roles:
1. A young man or woman describing how wonderful a parent is when there is a history of abuse by that parent.
2. A senior who has just been diagnosed with a life-threatening illness wondering why he or she is not sleeping at night.
3. An employee working in an organization that is downsizing who is concerned about frequent outbursts of anger.
4. An adolescent who has just lost a longtime family pet and does not want to go out with friends.

Observer role Operate the equipment and complete the Practice Interview Checklist (see Table 11.1).

Table 11.1

Practice Interview Checklist

Activity	Interviewer response[a]				
	1	2	3	4	5
Appropriate nonverbal behavior					
On topic					
Interpretation					
Other appropriate interviewing response					
Effective response[b]					

[a] Check the activities that characterize each response.

[b] Indicate the degree of effectiveness on a scale from 1 to 5 in which 1 is ineffective and 5 is very effective.

Postinterview discussion Using the taped interview and the Practice Interview Checklist data, the participants should discuss the performance of the interviewer. All participants can learn from this discussion, which should be nonjudgmental, focus on positive as well as less effective responses, and stimulate improved performance for all.

11.2

Activity Unit

Using the tape of the interview produced in Activity Unit 11.1 or the tape of another practice interview, enter the original interview responses below. Then formulate a better response to replace each original one in the space provided.

Original response 1 _____

_____

_____

Better response _____

_____

_____

Original response 2 _____

_____

_____

Better response _____

_____

_____

Original response 3 _____

_____

_____

Better response _____

_____

_____

Original response 4 _____

_____

_____

Better response _____

_____

_____

Original response 5 _____

_____

_____

Better response _____

_____

_____

Cultural Considerations

1. Before using interpreting responses, identify whether the cultural group with which the client is affiliated engages in self-exploration.
2. If used, ensure that the content of the interpreting response is relevant to the client's cultural affiliation.

InfoTrac® College Edition Keyword Search Terms

Biases AND Counseling
Interpretation AND Counseling

Chapter Twelve

Structuring for Information and Action

*T*his chapter is intended to help you perfect the skill of structuring. The chapter begins with a review of the skill of structuring for information, followed by material that helps you master the skill of structuring for action. After completing this chapter, you should be able to:

1. State the relationship objectives for each phase of the interviewing process.
2. State the action objectives for each phase of the interviewing process.
3. Decide when to make a structuring response.
4. Make structuring responses when they are required.

Structuring involves the ability to organize and pace a relationship with a client from its beginning to its conclusion. Structuring gives purpose and direction to the interview and enables the client and the interviewer to define and work toward specific outcomes. In each phase, the interviewer has a relationship objective and an action objective. To use the skill of structuring effectively, you must become familiar with the objectives listed in Table 12.1.

Table 12.1

An Overview of the Interviewing Phases Associated with Structuring for Information and Action

Interviewing goal	Relationship objective	Action objective
1. Exploration	Initiate a working relationship between the interviewer and the client.	Discover the client's problem(s).
2. Problem definition	Develop a more focused and facilitative relationship between the interviewer and the client.	Agree on a definition of the client's specific problem(s).
3. Problem priorization	Maintain a facilitative relationship between the interviewer and the client.	Assess the need for further action. If required, mutually agree on the order in which to consider the problems.
4. Goal-formulation	Maintain a facilitative relationship between the interviewer and the client. Use confrontation to help the client translate problems into goals.	Mutually redefine the problems as attainable goals. Each goal may need to be restated as a set of subgoals.
5. Preparation for action	Maintain a facilitative relationship between the interviewer and the client that acknowledges the client's responsibility.	Mutually generate action plans relevant to each goal and, based on the merits of each alternative, select the optimal plan.
6. Action	Maintain a facilitative relationship between the interviewer and the client by supporting or confronting as required.	Help the client implement the optimal action plan. Develop methods to maintain new behaviors in the absence of the interviewer–client relationship.
7. Termination	Conclude the interviewer–client relationship, leaving it open to reinitiation if required.	Terminate the interviewer–client relationship after the stated goals have been attained.

In the preceding chapters, you've been exposed to a variety of basic interviewing skills. As an interviewer, you will be most effective when you are able to combine and organize these skills in a purposeful manner—when you are able to structure your relationships with clients.

An effective interviewer uses the skill of structuring throughout the interviewer–client relationship. Because of the variety of decisions and problems that people will present to you, you must remain *flexible* in using this skill.

Occasionally, some phases of structuring may be unnecessary. On other occasions, each phase will be essential and will extend over a considerable period. In some cases, you may find that two phases blend together into a natural whole.

This chapter follows a series of interviews between a male student and a counselor at a student drop-in center. Go to 12.1.

12.1

The first phase of the interviewing process is the exploration phase, which initiates the interviewing process. Frequently, both the interviewer and the client are tense at the beginning of an initial interview. As the interviewer, it is your responsibility to be aware of this and to structure the opening of the interview to relax the client. Maintain appropriate eye contact and manner, make clear introductions, and if necessary, indicate seating arrangements with due regard to the client's culture. An open question frequently provides appropriate structure for the client to begin the discussion.

The interviewer has met the client and walked with him to an office.

Choose the best response to open the interview.

INTERVIEWER: (smiling, with appropriate eye contact, and directing the client to a chair) I haven't much time. What's your problem? Be as specific as you can. (Go to 12.2)

INTERVIEWER: (laughing nervously, with limited eye contact, and gesturing) I hope you really have a meaningful problem. Come and find a chair. What is it you want? (Go to 12.3)

INTERVIEWER: (smiling, with appropriate eye contact, and directing the client to a chair) You requested an appointment to discuss a problem. Can you tell me what you are concerned about? (Go to 12.4)

12.2   Your answer: *(smiling, with appropriate eye contact, and directing the client to a chair) I haven't much time. What's your problem? Be as specific as you can.*

Your response does provide structure for the client, but the structure suggests that you lack interest in what the client has to say. At this stage of the interview, the client requires a structure that will encourage rather than stop open discussion. Return to 12.1 and try again.

12.3   Your answer: *(laughing nervously, with limited eye contact, and gesturing) I hope you really have a meaningful problem. Come and find a chair. What is it you want?*

This response contains little structure to initiate the interview. The nervous laughter, poor eye contact, and nonspecific gesturing indicate your own discomfort with the interview. This behavior and your comments can only add to the client's discomfort. As a result, he may have difficulty responding to your brief open question. Return to 12.1 and try again.

12.4   Your answer: *(smiling, with appropriate eye contact, and directing the client to a chair) You requested an appointment to discuss a problem. Can you tell me what you are concerned about?*

Correct. This response supports the client in his request for an interview, facilitates open discussion, and structures the initial direction that the interview is to take. Go to 12.5.

12.5

Throughout the exploration phase, the interviewer employs listening and reflecting skills to demonstrate that he or she is listening and to help the client discuss problems openly. The first action objective in this phase is to gain an overview of the client's problem so the interviewer and the client can explore the situation together. As each problem is identified, the interviewer structures the interview to facilitate the emergence of any further problems.

The interviewer and the client have spent several minutes discussing the client's concerns about choosing a career.

CLIENT: One of the most pressing concerns I have is deciding whether I should go into dentistry or medicine.

Choose the most appropriate response.

INTERVIEWER: What does your father do? (Go to 12.6)

INTERVIEWER: A major concern for you is choosing the right profession. What are some of your other concerns? (Go to 12.7)

INTERVIEWER: That's a hard decision. I don't know which one I would choose. (Go to 12.8)

12.6   Your answer: *What does your father do?*

There was no indication in the client's statement that his father's profession is relevant to his decision. During this part of the exploration phase, you need to establish rapport with the client and try to understand the nature of his problems by employing listening skills and remaining on topic. Return to 12.5 and try again.

12.7   Your answer: *A major concern for you is choosing the right profession. What are some of your other concerns?*

Correct. During this part of the exploration phase, you need to establish rapport with the client and develop an initial understanding of his problems. By reflecting content and using an open question you've shown that you're listening to the client and have encouraged him to discuss his problems with you. Go to 12.9.

12.8   Your answer: *That's a hard decision. I don't know which one I would choose.*

This self-disclosure won't help facilitate an open relationship with the client, and it won't help him with the decision he has to make. Appropriate structuring at this point requires a response that indicates you have heard what the client has said and are willing to listen to his other concerns. Return to 12.5 and try again.

12.9

Once the range of the client's problems has been determined, the interviewer structures the interview from the exploration phase to the problem definition phase. In this phase, the interaction between the client and the interviewer becomes more focused. The action objective is to explore the client's problem areas until you both reach an accurate and concrete definition of each problem. To *achieve this goal, you may find it necessary to use action-oriented skills in addition to listening and reflecting skills.*

At the end of the exploration phase, the interviewer and the client determined that the client had difficulty deciding what career to enter, where to live, and how to obtain dates. The client alluded to other problems but preferred not to discuss them until later.

Choose the response that will initiate the problem definition phase.

INTERVIEWER: As you indicated earlier, you have a number of problems. Perhaps we could talk about these in detail and come up with a precise definition of each one. How does that sound to you? (Go to 12.10)

INTERVIEWER: Now, let's see. . . . You've talked about deciding between medicine and dentistry. That's a big decision. And you've indicated that you'd like to get away from your family. And then you mentioned dating. And perhaps you have other problems. I think we should solve the career problem first. Maybe you should go over and see the dean of medicine this afternoon. (Go to 12.11)

INTERVIEWER: About those other problems you think you have. . . . I know it's hard to talk about them, but before we can move on, you're going to have to tell me about them. (Go to 12.12)

12.10 Your answer: *As you indicated earlier, you have a number of problems. Perhaps we could talk about these in detail and come up with a precise definition of each one. How does that sound to you?*

Correct. This response orients the client toward the next phase of the interview. Go to 12.13.

12.11 Your answer: *Now, let's see. . . . You've talked about deciding between medicine and dentistry. That's a big decision. And you've indicated that you'd like to get away from your family. And then you mentioned dating. And perhaps you have other problems. I think we should solve the career problem first. Maybe you should go over and see the dean of medicine this afternoon.*

In this phase, you should help the client explore and define specific problems. This response omits exploration of any specific problem and offers a possibly inappropriate solution. Return to 12.9 and try again.

12.12 Your answer: *About those other problems you think you have. . . . I know it's hard to talk about them, but before we can move on, you're going to have to tell me about them.*

The client has already said that discussion of his other problems isn't warranted at this time. Therefore, you should move on to the next phase of the structuring process—the problem definition phase. Return to 12.9 and try again.

12.13

In the problem definition phase, the interviewer and the client explore problems until they agree on concrete definitions for each one. Usually, a concrete definition takes the form of a specific behavior that is problematic.

During the problem definition phase, the interviewer must be able to recognize the point at which the client formulates a precise and concrete definition of a problem. In the course of this interview, the client has formulated a definition of his dating problem.

Choose the comment that best defines the problem.

CLIENT: I guess I've got a poor self-concept that gets in my way when I'm out with a woman. (Go to 12.14)

CLIENT: I guess it's my parents' fault. They've never let me go out on dates. (Go to 12.15)

CLIENT: I guess I've just never learned how to arrange a date or what to do if I get one. (Go to 12.16)

12.14 Your answer: *I guess I've got a poor self-concept that gets in my way when I'm out with a woman.* To help the client, you must know specifically how his self-concept affects his behavior. In this situation, further exploration of the problem is needed. Return to 12.13 and try again,

---

12.15 Your answer: *I guess it's my parents' fault. They've never let me go out on dates.* This may be one aspect of the client's problem, but it isn't a precise definition of the problem in concrete terms. You need to encourage exploration beyond this point to reach a precise definition of the client's problem. Return to 12.13 and try again.

12.16 Your answer: *I guess I've just never learned how to arrange a date or what to do if I get one.* Correct. This is a precise definition of the client's problem. After you have helped him define his other problems, move on to the problem priorization phase of the interview. Go to 12.17.

12.17   *The third phase in the interviewing process is the problem priorization phase. During this phase, the interviewer continues to maintain a facilitative relationship with the client. The first action objective of the problem priorization phase is to determine the need for further involvement. The interviewer and the client must agree on this decision. In this case, the client now has an overview of his specific problems.*

CLIENT: Let me see. . . . I don't have enough information to decide what courses to take. I can't tell my parents I wish to leave home, and I've just never learned how to date.

Choose the response that will initiate the problem priorization phase.

INTERVIEWER: You've just identified three areas of concern. To what extent are you willing to work on developing some solutions to these problems? (Go to 12.18)

INTERVIEWER: You've just identified three areas of concern. We need to develop a solution to each one. (Go to 12.19)

INTERVIEWER: You've just identified three areas of concern, but I think they are things you can work on yourself. Let's close the interview at this point. It was a good idea for you to check out whether or not you needed help. (Go to 12.20)

12.18 Your answer: *You've just identified three areas of concern. To what extent are you willing to work on developing some solutions to these problems?*

Correct. This brief paraphrase, followed by an open question, is the appropriate structuring response at this point. The open question initiates a discussion that will help you and the client reach a decision concerning further action. Go to 12.21.

12.19 Your answer: *You've just identified three areas of concern. We need to develop a solution to each one.*

This response is premature. Before moving ahead, you and the client must agree on the need for further involvement. Return to 12.17 and try again.

12.20 Your answer: *You've just identified three areas of concern, but I think they are things you can work on yourself. Let's close the interview at this point. It was a good idea for you to check out whether or not you needed help.*

You've made a unilateral decision to terminate the interview; however, you and the client must reach a *mutual* decision concerning further involvement. Return to 12.17 and try again.

12.21

The second action objective in the problem priorization phase is to determine the order in which to consider the client's problems. The interdependence of problems may suggest a logical sequence. If a logical sequence isn't apparent and the client has no preference, move from the least severe problem to the most severe.

You and the client have agreed to discuss the following specific problems: his failure to obtain sufficient information on which to base his decision concerning his course of study, his inability to communicate to his parents his desire to leave home and attend college elsewhere, and his inability to arrange a date.

Choose the response that will continue the problem priorization phase of the interview.

INTERVIEWER: Now that we have decided to work together on your problems, we need to consider the order in which to deal with them. Your parents seem to have an important role in all of this. Before we go any further, you should tell them about your desire to leave home. (Go to 12.22)

INTERVIEWER: First, I want you to go home and draw up a list of all the women you know. Then, beside each name, write the reason for your inability to ask her for a date. (Go to 12.23)

INTERVIEWER: Now that we have decided to work together on your problems, we need to consider the order in which to deal with them. It seems that you're very concerned about choosing a career. It might be appropriate to deal with that first. What do you think? (Go to 12.24)

12.22 Your answer: *Now that we have decided to work together on your problems, we need to consider the order in which to deal with them. Your parents seem to have an important role in* *all of this. Before we go any further, you should tell them about your desire to leave home.*

You've imposed your priorities on the client and moved into problem solving much too quickly. The client and the interviewer must agree on the ranking of the client's problems. Moreover, you have shifted the relationship from an open, facilitative one to a closed, authoritarian one. Return to 12.21 and try again.

12.23 Your answer: *First, I want you to go home and draw up a list of all the women you know. Then, beside each name, write the reason for your inability to ask her for a date.*

During this phase of the interview, you need to help the client rank his concerns according to their priority. The client and the interviewer must agree on the ranking of the concerns. You've assumed that the dating problem supersedes all others. Moreover, you've failed to tell the client what he might expect during this phase of the interview. Return to 12.21 and try again.

12.24 Your answer: *Now that we have decided to work together on your problems, we need to consider the order in which to deal with them. It seems that you're very concerned about choosing a career. It might be appropriate to deal with that first. What do you think?*

Correct. This response explains the problem priorization phase of the interview and initiates the ordering process in a way that invites the client's input. Go to 12.25.

12.25

Up to this point, you've helped the client explore his concerns (exploration phase), define his concerns in concrete terms (problem definition phase), and determine the need for further involvement and the order in which these concerns should be considered (problem priorization phase). The remaining phases relate to structuring for action and involve setting goals, planning action, implementing programs, and terminating the interviewer–client relationship. To illustrate these phases clearly, we have focused the remaining segments of this program on only one of the client's concerns. Go to 12.26.

12.26

After you and the client have determined the order in which to deal with specific concerns, you move to the goal-formulation phase of structuring. The client's concern or problem, as stated in the problem definition phase, defines the present situation. The aim of goal formulation is to redefine this problem in terms of an alternative situation that is both realistic and acceptable to the client. Goals fall on a continuum from remediation to growth. At the conclusion of the problem priorization phase of this interview, the client makes the following statement.

CLIENT: Choosing a career is a major concern of mine, but I've got a whole year to work that out. The thing that's really getting me down is never having a date. All my friends get dates, but I just can't seem to get one.

INTERVIEWER: From what you've said, it sounds as though we should work on that first. What about that?

CLIENT: I'd be happy if we did that.

Choose the response that will best structure the goal-formulation phase.

INTERVIEWER: Maybe you'd better tell me about this dating problem again. I'm not sure I understand it completely. Have you been on any dates recently? (Go to 12.27)

INTERVIEWER: We've discussed your dating problem. Now we need to use that information to develop a goal that you're willing to work toward. Could you describe a goal that you'd like to reach? (Go to 12.28)

INTERVIEWER: Before I see you again, I want you to do three things. Spend 5 minutes chatting with a woman after each of your classes. Have lunch with a mixed group every day. And invite a woman for coffee at least once during the next week. (Go to 12.29)

12.27 Your answer: *Maybe you'd better tell me about this dating problem again. I'm not sure I understand it completely. Have you been on any dates recently?*

You've already collected the information relevant to this particular problem. When you ask a client to repeat information, you waste time and do little to facilitate your relationship or help the client set appropriate goals. Return to 12.26 and try again.

12.28 Your answer: *We've discussed your dating problem. Now we need to use that information to develop a goal that you're willing to work toward. Could you describe a goal that you'd like to reach?*

Correct. You've invited the client to redefine his problem as an attainable goal. Go to 12.30.

12.29 Your answer: *Before I see you again, I want you to do three things. Spend 5 minutes chatting with a woman after each of your classes. Have lunch with a mixed group every day. And invite a woman for coffee at least once during the next week.*

This response violates the mutuality of the interviewing process. The client is more likely to work toward a goal that he has helped formulate and to which he aspires. Moreover, in making this response, you've skipped both the goal-formulation and the preparation for action phases of interviewing. All possibilities should be examined before an action plan is initiated. Return to 12.26 and try again.

12.30

A client may have difficulty translating a problem into an attainable goal. If this happens because the problem has been poorly defined, return to the problem definition phase and develop a clearer definition. However, you may find that the client can define the problem but seems unwilling or unable to translate it into an attainable goal. When this occurs, you may want to use the skill of confronting to focus on this unwillingness or inability.

CLIENT: I don't know about any goal. I don't think you understand. Maybe I'd better tell you more about it, but I'm not sure it's that much of a problem.

Choose the most appropriate response.

INTERVIEWER: It's much easier to talk about your problems than it is to solve them. Deciding on a goal is difficult. (Go to 12.31)

INTERVIEWER: I think your goal should be to go out with women until you have a list of four or five who go out with you regularly. (Go to 12.32)

INTERVIEWER: You don't seem to be too sure of what you want. Maybe we should talk about it some more. Where do you want to start? (Go to 12.33)

12.31 Your answer: *It's much easier to talk about your problems than it is to solve them. Deciding on a goal is difficult.*
Correct. By confronting the client with the discrepancy between his stated desire to act and his apparent unwillingness to act, you encourage him to discuss and set appropriate, realistic goals. Go to 12.34.

12.32 Your answer: *I think your goal should be to go out with women until you have a list of four or five who go out with you regularly.*
Your response should help the client explore his unwillingness to formulate a goal. You've ignored his unwillingness and suggested a goal that may or may not be realistic or appropriate. Although the client may cooperate with you temporarily, it is doubtful that he will persist if he is having difficulty moving into the next phase of interviewing. Return to 12.30 and try again.

12.33 Your answer: *You don't seem to be too sure of what you want. Maybe we should talk about it some more. Where do you want to start?*
You've already explored the problem with the client and developed a definition of it. It is unlikely that you will alter the client's resistance to establishing a goal by covering this information again. You need to confront him with his behavior to help him explore and clarify its meaning and work toward the establishment of a goal. Return to 12.30 and try again.

12.34

Goals should be formulated in concrete terms. To accomplish this, you may need to redefine a general goal as a series of specific subgoals, or steps, that lead to the attainment of the general goal.

CLIENT: I hadn't looked at it that way. I guess I'm a little scared to think of going out with women. But I know I've got to do something about it.

INTERVIEWER: Let's talk about the way you'd like things to be. How would you act toward women if you didn't have this problem?

CLIENT: I'd like to be able to ask women out so that I could have some fun. My friends know women, and they all do things together.

Choose the response that will continue the goal-formation phase of this interview.

INTERVIEWER: I think that's a reasonable goal. Perhaps we can move on now to identify some ways of doing that. (Go to 12.35)

INTERVIEWER: It seems that you envy your friends and what they do. Could you tell me about some of those things? (Go to 12.36)

INTERVIEWER: That's a reasonable goal. What might be the first step in learning to ask women out? (Go to 12.37)

12.35 Your answer: *I think that's a reasonable goal. Perhaps we can move on now to identify some ways of doing that.*

You're moving too quickly. This structuring response is inappropriate because the client's goal is too vague to act as a reasonable focus for preparation for action or action initiation. Your response should help the client define his goal in more concrete terms. Return to 12.34 and try again.

12.36 Your answer: *It seems that you envy your friends and what they do. Could you tell me about some of those things?*

This response may be appropriate during some other phase of the interview, but not when you're helping the client define goals. You've changed the topic and asked an open question that is irrelevant to the solution of the client's problem. Return to 12.34 and try again.

12.37 Your answer: *That's a reasonable goal. What might be the first step in learning to ask women out?* Correct. This response encourages the client to define his general goal in a series of concrete, attainable subgoals. Go to 12.38.

12.38

After a specific goal or set of subgoals has been identified, the interviewer and the client need to agree on an action plan that will enable the client to attain the goal or subgoals. It is important for the client to be responsible for the formulation of the action plan. The first step is to generate a number of plans that are relevant to the client's goal. Although the client should be mainly responsible for developing action plans, the interviewer may need to contribute possible alternatives.

The interviewer and the client have now agreed on a number of subgoals. First of all, the client is to have coffee with a woman. For purposes of illustration, this subgoal alone will be dealt with.

Choose the response that will initiate the preparation for action phase of this interview.

INTERVIEWER: I think it's important that you have coffee with a woman you don't know. That seems to be a good way to start. (Go to 12.39)

INTERVIEWER: I suggest that you ask a woman to have coffee on the spur of the moment and go with her right away. That way, you'll keep anxiety to a minimum. (Go to 12.40)

INTERVIEWER: We need to think about how you might accomplish this goal. What ideas do you have? (Go to 12.41)

12.39 Your answer: *I think it's important that you have coffee with a woman you don't know. That seems to be a good way to start.*

It is wise to give the client an opportunity to formulate his own plan of action. This response limits his responsibility. At this point in the preparation for action phase, your response should encourage the client to participate in the generation of possible alternatives. Return to 12.38 and try again.

12.40 Your answer: *I suggest that you ask a woman to have coffee on the spur of the moment and go with her right away. That way, you'll keep anxiety to a minimum.*

You've taken full responsibility for generating the action plan. Action plans are most likely to be successful when clients generate and explore their own alternative routes to reaching their goals. Return to 12.38 and try again.

12.41 Your answer: *We need to think about how you might accomplish this goal. What ideas do you have?*

Correct. You've encouraged the client to focus on possible action plans for the goal under discussion and to take responsibility for generating these plans. Go to 12.42.

12.42   *After relevant action plans have been established, the advantages and disadvantages of each one are explored, and an optimal plan is selected.*

The interviewer and the client have explored four relevant plans of action.

Choose the response that will further structure the preparation for action phase of this interview.

INTERVIEWER: We've decided on four possible methods by which you can attain your goal. You could choose someone ahead of time and arrange a meeting over the phone the night before. You could ask a woman you know through your friends. You could casually ask someone in one of your classes as the class ends. Or you could ask someone who never seems to go out. We've discussed the consequences of each of these plans and how the plans themselves fit in with your values. From your perspective, which alternative seems most appropriate? (Go to 12.43)

INTERVIEWER: You said you would find it difficult to ask a woman to have coffee with you if you'd met her through your friends. You seem to think that doing this would hurt your relationship with your friends. That seems out of place in this day and age. Maybe we should think about clarifying your values before we go any further. (Go to 12.44)

INTERVIEWER: We've decided on four possible methods by which you can attain your goal. We've discussed the consequences of each of these plans and how the plans themselves fit in with your values. From my experience, and from what you've said, I think that you should ask someone ahead of time and set up a date over the phone the night before. (Go to 12.45)

12.43 Your answer: *We've decided on four possible methods by which you can attain your goal. You could choose someone ahead of time and arrange a meeting over the phone the night before. You could ask a woman you know through your friends. You could casually ask* *someone in one of your classes as the class ends. Or you could ask someone who never seems to go out. We've discussed the consequences of each of these plans and how the plans themselves fit in with your values. From your perspective, which alternative seems most appropriate?*

Correct. In this structuring response, you've briefly summarized the most promising plans of action and invited the client to choose the plan with which he feels most comfortable. Go to 12.46.

12.44 Your answer: *You said you would find it difficult to ask a woman to have coffee with you if you'd met her through your friends. You seem to think that doing this would hurt your relationship with your friends. That seems out of place in this day and age. Maybe we should think about clarifying your values before we go any further.*

Action plans must be formulated with the client's values in mind. However, if the client's values are the cause of some problem, this problem should be defined and considered at another time. At this point, you should consider alternatives that are open to the client. Return to 12.42 and try again.

12.45 Your answer: *We've decided on four possible methods by which you can attain your goal. We've discussed the consequences of each of these plans and how the plans themselves fit in with your values. From my experience, and from what you've said, I think that you should ask someone ahead of time and set up a date over the phone the night before.*

This response renders the discussion of alternatives and their consequences useless. The purpose of discussing alternatives is to expose clients to a number of possible methods by which goals can be attained and to permit them to select the method with which they are most comfortable. In attempting to impose a solution, you ignore the standards and values of the client. Return to 12.42 and try again.

12.46

During the action phase of interviewing, the interviewer helps the client implement the action plan that has been selected. After the plan has been implemented, the interviewer and the client should evaluate its success according to the criteria established for attaining the goal. Based on this evaluation, they may agree to modify the action plan or to select a more appropriate one. If the plan is successful, the client and the interviewer need to consider ways by which the client's new behaviors can be maintained without the support of the interviewing process.

During the previous session, the client decided to ask a woman to have coffee with him at the conclusion of one of his classes. Now he is back to report on the results of this plan of action.

CLIENT: Well . . . I did it. I asked the most attractive woman in my class to have coffee with me. I managed to do it. But she said she didn't have time. Then I asked her to go out with me the next day, and she made some other excuse.

Choose the most appropriate response.

INTERVIEWER: Sounds like it was a disaster. I think we'd better go back and select another plan. (Go to 12.47)

INTERVIEWER: Sounds like asking her went well, but the rest didn't go so well. What could you do next time to achieve your goal of asking a woman to have coffee with you? (Go to 12.48)

INTERVIEWER: Sounds like asking her went well. But did you really expect to succeed with the most attractive woman in your class? (Go to 12.49)

12.47 Your answer: *Sounds like it was a disaster. I think we'd better go back and select another plan.*
You need to discuss the client's failure to achieve his goal in order to help him decide whether abandoning the goal is appropriate. A slight adjustment in the plan might help him attain his goal. The client should help decide the characteristics of the plan needed. Return to 12.46 and try again.

12.48 Your answer: *Sounds like asking her went well, but the rest didn't go so well. What could you do next time to achieve your goal of asking a woman to have coffee with you?*
Correct. This response acknowledges the client's partial success and encourages him to consider what he can do to achieve his goal. Go to 12.50.

12.49 Your answer: *Sounds like asking her went well. But did you really expect to succeed with the most attractive woman in your class?*
The initial part of your response is supportive, but then you belittle the client's attempt and thus remove the effect of your supportive statement. The client's choice of a potential companion may have contributed to his lack of success. As the interviewer, you should help him recognize this fact. Return to 12.46 and try again.

12.50   *Clients may fail to cooperate with you during the action phase of interviewing for a number of reasons: they may experience a conflict concerning the selected goal, they may lack the skills they need to attain the goal, they may not want to change their behavior, or they may experience problems in the client–interviewer relationship. When clients fail to cooperate, you need to help them confront and remove the impediments to attaining their goals.*
Several weeks have passed, and the client is regularly having coffee with women and talking with them on the phone. However, he has been unable to ask a woman for a date, despite several attempts to initiate an action plan for this goal.

Choose the most appropriate comment.

INTERVIEWER: You seem to be having a great deal of difficulty reaching this goal. Perhaps we should work on something else for a while. (Go to 12.51)

INTERVIEWER: You want to ask a woman for a date, but you've been unable to reach this goal. Could you help me to understand your difficulty? (Go to 12.52)

INTERVIEWER: Perhaps your difficulty in attaining your goal is caused by a fear of women. What do you think has contributed to that fear? (Go to 12.53)

12.51 Your answer: *You seem to be having a great deal of difficulty reaching this goal. Perhaps we should work on something else for a while.*
When a client experiences difficulty in carrying out a plan of action, you need

to help him or her confront and resolve the difficulty. As an interviewer, it is your responsibility to help the client identify and explore difficulties rather than avoid them. Return to 12.50 and try again.

12.52 Your answer: *You want to ask a woman for a date, but you've been unable to reach this goal. Could you help me to understand your difficulty?*
Correct. This response confronts the client with his failure to reach the identified goal and invites him to explore the reasons for his failure. Go to 12.54.

12.53 Your answer: *Perhaps your difficulty in attaining your goal is caused by a fear of women. What do you think has contributed to that fear?*
This statement is an assumption that is not supported by the information you have. As indicated in the introduction to this set of frames, a client may resist goal attainment for a number of reasons. You need to confront the client with his apparent resistance rather than guess at the reason behind it. By using confrontation at this point, you'll invite the client to discuss the reasons for his resistance. Return to 12.50 and try again.

12.54

In the termination phase of structuring, the interviewer and the client evaluate the degree to which the client has reached the overall goal. If the client has been unsuccessful, it may be necessary to formulate an alternative action plan or an alternative goal. If the client has successfully accomplished the goal, it may be necessary to move on to other goals. During the termination phase, you may want to review your relationship so you and the client can gain insight into the client's relationships with others. While concluding the interview, leave the relationship open so the client will feel free to contact you again, if necessary.

Three months have passed since the preceding interview. The client dates regularly, has decided to study dentistry, and has told his parents that he is going away to college. They've accepted this idea very well.

Choose the response that will initiate the termination phase.

INTERVIEWER: You've accomplished a great deal in the past 3 months. You've developed the ability to go out with women. You've decided to study dentistry, and you've reached a point of agreement with your parents concerning your plans for the future. I don't think we need to meet anymore. I hope that you're successful in dentistry. (Go to 12.55)

INTERVIEWER: You've accomplished a great deal in the past 3 months. You've developed the ability to go out with women. You've decided to study dentistry, and you've reached a point of agreement with your parents concerning your plans for the future. You must have problems we haven't discussed yet. Could we discuss them now? (Go to 12.56)

INTERVIEWER: You've accomplished a great deal in the past 3 months. You've developed the ability to go out with women. You've decided to study dentistry, and you've reached a point of agreement with your parents concerning your plans for the future. Now that we've reached this point, I wonder if there are any other concerns you would like to discuss. (Go to 12.57)

12.55 Your answer: *You've accomplished a great deal in the past 3 months. You've developed the ability to go out with women. You've decided to study dentistry, and you've reached a point of agreement with your parents concerning your plans for the future. I don't think we need to meet anymore. I hope that you're successful in dentistry.*

Before you terminate the interview, you need to determine whether or not there are other concerns the client wants to deal with. Moreover, when you conclude the interview, you should make it clear that the client can contact you again, if necessary. Return to 12.54 and try again.

12.56 Your answer: *You've accomplished a great deal in the past 3 months. You've developed the ability to go out with women. You've decided to study dentistry, and you've reached a point of agreement with your parents concerning your plans for the future. You must have problems we haven't discussed yet. Could we discuss them now?*

The client has spent 3 months exploring and resolving his problems. At this and other stages of interviewing, your relationship with the client is best facilitated by open rather than directive discussion. Your assumption that the client has other problems is not appropriate at this time. Return to 12.54 and try again.

12.57 Your answer: *You've accomplished a great deal in the past 3 months. You've developed the ability to go out with women. You've decided to study dentistry, and you've reached a point of agreement with your parents concerning your plans for the future. Now that we've reached this point, I wonder if there are any other concerns you would like to discuss.*

Correct. This response indicates that it is appropriate to discuss any other problems the client may have. This brief check is an important step in the termination phase. Go to 12.58.

12.58

Remember that you must remain flexible when using the skill of structuring. The frames in this chapter were designed to illustrate the important phases of structuring. Whether or not you use each phase with a client will be determined by the nature of the problem. At times, it may be necessary to repeat the early phases of structuring. Go to 12.59.

12.59

Review Questions

Indicate whether each of the following statements is true or false.

1. During the problem definition phase, an accurate definition of the client's problems is developed.
2. It is important that the interviewer offers solutions to the client's problems as soon as possible.
3. Interviewers should always encourage clients to deal with their easiest problem first.
4. Once the client's problems have been defined, an action plan for each one should be developed.

5. Difficulties in goal formulation often result from poor definition of problems and from the client's resistance to change.
6. During the preparation for action phase, the client should retain maximum responsibility.
7. The most appropriate action plan is the one that is most appealing to the client.
8. The interviewer can be the best judge of when the interview should terminate.
9. Every client should progress through each interviewing phase.

Review Answers

1. *True.* If you answered *false,* go to 12.13 and review.
2. *False.* If you answered *true,* go to 12.5 and review.
3. *False.* If you answered *true,* go to 12.21 and review.
4. *False.* If you answered *true,* go to 12.25 and review.
5. *True.* If you answered *false,* go to 12.30 and review.
6. *True.* If you answered *false,* go to 12.38 and review.
7. *False.* If you answered *true,* go to 12.42 and review.
8. *False.* If you answered *true,* go to 12.54 and review.
9. *False.* If you answered *true,* go to 12.58 and review.

If six or more of your responses were incorrect, you should return to 12.1 and review the material in this chapter.

Points to Remember about Structuring

1. USE STRUCTURING TO:
   a. Initiate or conclude an interviewing phase.
   b. Facilitate attainment of the objective for a given phase.
   c. Provide the client with information concerning the direction of the interview.
2. WHEN STRUCTURING:
   a. Initiate a new phase by stating its purpose, and then determining the client's willingness to participate (check statement).
   b. Use previously acquired skills to facilitate attainment of the objective for a given phase.
   c. Conclude a phase by reviewing what has been accomplished and as certaining whether the client agrees that the objectives have been met.

12.1

Activity Unit

Working with two others who have also read this chapter, role play short interviews alternating the client, interviewer, and observer roles. Record each brief interview on audio- or videotape. A brief interview, covering between 10 and 15 interviewer responses, is sufficient for this exercise. You may wish to follow the method, used in this chapter, in which the client or interviewer narrates what has occurred between responses, so that the interviewer can practice the next type of structuring response.

Interviewer role In the set of brief interviews, practice structuring responses.

Client role Be cooperative and facilitate the progress of the interviews so the interviewer can practice the skills of structuring. Clients may want to role play the problem of a friend or family member or to enact one of the following roles:

1. A person who must change occupations but is fearful of engaging in the required retraining program.
2. A husband or wife who has just separated and is fearful and confused about the future.
3. An individual who has just been informed of the chronic illness of her or his remaining parent. Some decision is required concerning the support and management of the parent.
4. A student who has concluded that he or she is pursuing an inappropriate program of study. The student is unsure of what alternate course to follow.

Observer role Operate the equipment and complete the Practice Interview Checklist in Table 12.2 at the end of the chapter.

Postinterview discussion Using the taped interviews and the Practice Interview Checklist data, the participants should discuss the performance of the interviewer. All participants can learn from this discussion, which should be nonjudgmental, focus on positive as well as less effective responses, and stimulate improved performance for all.

12.2

Activity Unit

Using the tape of the interview produced in Activity Unit 12.1 or the tape of another practice interview, enter the original interview responses that are least effective below. Then formulate a better response to replace each original one in the space provided.

Original response 1 _____

_____

_____

Better response _____

_____

Table 12.2
Practice Interview Checklist

Activity	Practice interview 1		Practice interview 2	
	Yes	No	Yes	No
1. Structured the beginning of the interview.				
a. maintained appropriate eye contact.				
b. projected relaxed, professional manner.				
c. provided clear introductions.				
d. used open inquiry.				
e. other _____.				
2. Attained the goals of the exploration phase.				
a. initiated a working relationship.				
b. discovered the client's problems.				
c. other _____.				
3. Achieved transition from exploration to the problem definition phase.				
a. summarized exploration.				
b. indicated aims of the problem definition.				
c. used check statement.				
4. Attained the goals of the problem definition phase.				
a. developed a more focused and facilitative relationship.				
b. obtained specific definitions of problems.				
5. Achieved transition from problem priorization phase.				
a. summarized problem definition phase.				
b. indicated aims of problem priorization phase.				
c. used check statement.				
6. Attained the goals of the problem priorization phase.				
a. maintained a facilitative relationship.				
b. assessed the need for further action.				
c. developed priority of problems.				
7. Achieved transition from problem priorization phase to goal-formulation phase.				
a. summarized problem priorization.				
b. indicated aims of goal-formulation phase.				
c. used check statement.				

Table 12.2

Practice Interview Checklist

Activity	Practice interview 1		Practice interview 2	
	Yes	No	Yes	No
8. Attained the goals of goal-formulation phase.				
a. maintained a facilitative relationship.				
b. redefined problems as goals.				
c. developed subgoals if required.				
9. Achieved transition from goal-formulation phase to action preparation for phase.				
a. summarized goal-formulation phase.				
b. indicated aims of preparation for action phase.				
c. used check statement.				
10. Attained goals of preparation for action phase.				
a. maintained facilitative relationship.				
b. acknowledged the client's responsibility in formulating action plans.				
c. generated action plans for each goal.				
d. discussed merits of each action plan.				
e. selected optimal plan.				
11. Achieved transition from preparation for action to action phase.				
a. summarized preparation for action phase.				
b. indicated aims of action phase.				
c. used check statement.				
12. Attained goals of action phase.				
a. maintained a facilitative relationship.				
b. implemented optimal action plan.				
c. planned for independent behaviors.				
13. Structured termination.				
a. supported the client in termination.				
b. negotiated termination.				
c. left the relationship open to reinitiation, if required.				

Original response 2 _____

_____

_____

Better response _____

_____

_____

Original response 3 _____

_____

_____

Better response _____

_____

_____

Original response 4 _____

_____

_____

Better response _____

_____

_____

Original response 5 _____

_____

_____

Better response _____

_____

_____

Cultural Considerations

1. Effective structuring with clients from other cultures requires the interviewer to help the client define goals and develop action plans that are consistent with the life experiences and cultural values of the client.

2. When interviewing clients from some cultures, it is important that the interviewer be pragmatic, address immediate concerns, and take a more active role in the structuring process.

InfoTrac® College Edition Keyword Search Terms

Structuring AND Interviewing

Chapter Thirteen

Enlisting Cooperation

*T*his chapter is intended to help you enlist cooperation by using the skills you have learned. When beginning interviewers encounter lack of client cooperation they can become confused, angry, guilty, or depressed, losing direction in the interview. Often they respond in nonproductive ways, such as trying to placate the client, becoming impatient or hostile, engaging in a power struggle, or trying to distance themselves from the client. The skill of enlisting cooperation will assist interviewers to become aware of issues underlying lack of client cooperation, and will help interviewers to develop strategies for their resolution. Assisting clients to cooperate can be a simple or a highly complex task. Not all resistant clients will respond to an interviewer's attempts to enlist cooperation, no matter how skilled the interviewer. However, basic interviewing and action-oriented skills can be used effectively in assisting clients to become more comfortable and less resistant to the interviewing process. After completing this chapter, you should be able to:

1. Identify when a client is being resistant or uncooperative.
2. Identity the reason for client resistance.
3. Use the skills presented in this book to enlist cooperation.
4. Assist a client to move from one stage of change to another.

All theories of interviewing discuss the importance of enlisting cooperation. Lack of cooperation has been described by many terms, including *client resistance, reluctance, discomfort,* and *noncompliance. Resistance* is the term most commonly used to describe any client or interviewer behavior that interferes with or reduces the probability of a successful outcome to the interviewing process or to a given phase of the interviewing process. In managing resistance, the goal of the interviewer is to enlist or restore client cooperation.

Difficulties with client cooperation can occur for a variety of reasons and at any phase in the interviewing process. Although lack of client cooperation can occur at any phase, it is encountered most frequently during the development phase. Other phases in which clients are more likely to encounter difficulties in cooperating are the action and termination phases. Client lack of cooperation is expressed in many ways and for numerous reasons. Potential reasons for client resistance are shown in Table 13.1.

Sooner or later in the interviewing process, clients experience difficulty in participation. Identification of difficulties with cooperation is a matter of the individual judgment of the interviewer, the nature of the client's behavior, and the context of the behavior. The interviewer avoids premature judgment that a particular client behavior represents lack of cooperation in the interviewing process, but he or she is aware that repeated occurrences do suggest problems with cooperation. Effective interviewers are not surprised by lack of cooperation, nor do they react to it at a personal level. Rather, they use the skills included in this book to enlist the cooperation of clients in achieving the goals of the interviewing process.

The frames in this chapter are both independent and sequential in order to demonstrate a range of situations in which clients may be uncooperative. Included also is the use of this skill to assist clients to move from one interviewing phase or stage to the next. Go to 13.1.

13.1

Clients demonstrate their difficulty with the interviewing process in a number of ways. Interviewers can identify clients who are experiencing difficulties in cooperating by observing the amount of information given by the client. Some clients may react to an interview by limiting the amount of information communicated, whereas others may overwhelm the interviewer with information.

INTERVIEWER: (a high school counselor interviewing a 15-year-old who is persistently late and fails to complete assignments) You were saying that you hate school. Could you tell me what you hate about school?

Choose the response that demonstrates lack of cooperation caused by the client controlling the amount of information shared.

Table 13.1

Reasons for Client Resistance

Source of resistance	Reason for resistance
Client	
Ability	Lack of skills or knowledge required to participate
	Purpose of the interview not understood
Emotion	Shame associated with needing help or assistance
	Pessimistic orientation or negative thoughts
	Fear of the interviewing relationship
	Anxiety about discussion of personal information
	Anxiety about losing control over the nature and direction of discussion in the interview
	Fear of confronting feelings associated with the topic under discussion
	Threat to independence
	Importance of maintaining a rebellious stance
Motivation	Loyalty to others
	Unwilling to confront the disruption associated with change
	Unwilling to make the required change
	Lack of any motivation for meeting the interviewer
	Resentment of those (court, agency, and so on) requiring the interview being extended to the interviewer
	Interview felt to be unnecessary
	Difference between own goals and those of the system (for example, court, place of work, family)
Interviewer	Values (cultural or personal) do not match the values of the client
	Personal characteristics (age, race, gender, appearance, style of interviewing) of concern to the client
	Inability to provide the required level of client confidentiality
Environment	Not supportive of the desired change
	Supports not changing
	Cultural traditions not supportive of intervention by outsiders
	Cultural norms suggestive of seeking help as sign of weakness

CLIENT: My father says I don't need to be here anymore. He quit at 14, and he's done all right. (Go to 13.2)

CLIENT: I find school very hard. I try to do my work, but I never seem to get anywhere. (Go to 13.3)

CLIENT: I don't know. I just hate it. (Go to 13.4)

13.2   **Your answer:**   *My father says I don't need to be here anymore. He quit at 14, and he's done all right.*
Although the client has not responded with information that is directly associated with the interviewer's inquiry, there has been some attempt to cooperate with the interviewer by providing information related to a potential reason for difficulties at school. Return to 13.1 and try again.

13.3   **Your answer:**   *I find school very hard. I try to do my work, but I never seem to get anywhere.*
In this response, the client provides information that will permit the interview to move forward. It neither limits discussion nor overwhelms the interviewer with information. Return to 13.1 and try again.

13.4   **Your answer:**   *I don't know. I just hate it.*
Correct. The client shows reluctance to participate in the interview by failing to provide additional information in response to the interviewer's open inquiry. Go to 13.5.

13.5

Another way that clients show resistance in an interview is by restricting what is discussed. Clients may ramble, engage in repetitive, superficial, or intellectual discussion of selected topics, or continually present themselves as unable to move past their emotional reaction to the current focus of the interview.

INTERVIEWER: (a medical social worker interviewing a client about to be discharged) Your physician has referred you to me so that I can help plan your release from the hospital. My knowing whether you have family and friends is important to planning. Can you tell me about your home situation?

Choose the response that demonstrates lack of cooperation caused by the client's controlling the content of the interview.

CLIENT: My back is very sore today. I was awake all night. I couldn't sleep at all. The nurse wouldn't give me anything for it. I'm in a lot of pain right now, and I need to talk to my physician. (Go to 13.6)

CLIENT: (sits in silence, avoids eye contact, and fails to respond) (Go to 13.7)

CLIENT: I'm not sure how I'll manage. I live alone. There are lots of stairs in my house, and I live out of town. (Go to 13.8)

13.6   **Your answer:**   *My back is very sore today. I was awake all night. I couldn't sleep at all. The nurse wouldn't give me anything for it. I'm in a lot a pain right now, and I need to talk to my physician.*
Correct. Reluctance to cooperate in discussion of discharge plans, the content of the interviewer's inquiry, is shown by the client's avoidance of the topic and by the client's repetitive discussion of physical symptoms. Go to 13.9.

13.7   **Your answer:**   *(sits in silence, avoids eye contact, and fails to respond)*
The client is uncooperative but is limiting the amount of information rather than controlling the content of the interview. Return to 13.5 and try again.

13.8　**Your answer:**　*I'm not sure how I'll manage. I live alone. There are lots of stairs in my house, and I live out of town.*

The client is cooperative and is providing important information in response to the interviewer's open inquiry. Return to 13.5 and try again.

13.9

Clients may show lack of cooperation by the way they communicate information to the interviewer. Clients may react with boredom, respond seductively, flatter the interviewer, or seek arguments. They may blame others, engage in self-justification, discount the importance of a topic, or withhold important information until the end of the interview.

INTERVIEWER:　(a personnel consultant interviewing a prospective employee) I'd like to shift the focus of discussion now. Could you tell me about your last job?

Choose the response that demonstrates lack of cooperation caused by the client's manner of responding.

CLIENT:　My last job was with an oil company. They have oil holdings all over the world. In recent years, the cost of oil has increased dramatically and so have the profits. (Go to 13.10)

CLIENT:　My last job was with an oil company. I was a regional sales manager, supervising eight offices with 120 employees. As a result, I was on the road quite a lot. (Go to 13.11)

CLIENT:　I'm quite impressed with your interviewing style. You provide lots of support and ask good questions. I'd enjoy working with you. (Go to 13.12)

13.10　**Your answer:**　*My last job was with an oil company. They have oil holdings all over the world. In recent years, the cost of oil has increased dramatically and so have the profits.*

In this response, the client controls the content of the interview by talking about the company rather than the requested personal information. The failure to cooperate is achieved by controlling the content rather than by the style of the response. Return to 13.9 and try again.

13.11　**Your answer:**　*My last job was with an oil company. I was a regional sales manager, supervising eight offices with 120 employees. As a result, I was on the road quite a lot.*

In this response, the client has cooperated with the interviewer and shared information about the last job, which enables the interview to progress. Return to 13.9 and try again.

13.12　**Your answer:**　*I'm quite impressed with your interviewing style. You provide lots of support and ask good questions. I'd enjoy working with you.*

Correct. Instead of responding directly to the interviewer's inquiry, the client uses a particular manner of responding, flattery, to divert the interviewer from the information requested. Go to 13.13.

<table>
<tr><td>13.13</td><td></td></tr>
</table>

13.13

A final way that clients show lack of cooperation is by failing to keep the formal or informal contract with the interviewer. Clients may fail to attend interviews, arrive late, request additional time, or impose in other ways on the interviewer.

INTERVIEWER: (closing an interview with a client who has been late for this and other appointments) We have discussed and resolved your immediate concerns of this past week. Unfortunately, that's all the time we have for today.

Choose the response that demonstrates lack of cooperation caused by the client's failure to keep the contract established with the interviewer.

CLIENT: I'm sorry I was late. I know you have another appointment now. I'll be on time next week. My visitors will have gone by then. (Go to 13.14)

CLIENT: I'm sorry I was late. I can't leave yet. There are a number of things I need to talk about, and they can't wait until next week. Surely you can find some time for me. (Go to 13.15)

CLIENT: I'm sorry I was late. It was one of those days. I find it hard to be on time when I have visitors. I'll be more organized next week. (Go to 13.16)

13.14 Your answer: *I'm sorry I was late. I know you have another appointment now. I'll be on time next week. My visitors will have gone by then.*
This cooperative response shows appreciation for both the interviewer's position and the interviewing contract. Return to 13.13 and try again.

13.15 Your answer: *I'm sorry I was late. I can't leave yet. There are a number of things I need to talk about, and they can't wait until next week. Surely you can find some time for me.*
Correct. The client shows a lack of cooperation by failing to keep the formal or informal contract established with the interviewer concerning the timing of interviews. Go to 13.17.

13.16 Your answer: *I'm sorry I was late. It was one of those days. I find it hard to be on time when I have visitors. I'll be more organized next week.*
This cooperative response demonstrates the client's awareness of responsibility for keeping the rules of the interviewing agreement. Return to 13.13 and try again.

13.17

During the early phases of interviewing, it is not unusual for clients to experience difficulty with sharing information. Attempts to focus on the purpose of the interview will be unsuccessful until the interviewer has enlisted the client's cooperation by resolving the source of discomfort.

The next sequence of frames focuses on the early stage of an interview between a young African American female counselor and an older white male who has come to an employment center for assistance in finding work.

CLIENT: (somewhat angry) I've been out of work for nearly a year, and this is my last hope of finding a job. I've tried everything else. When I finally make it here, I am assigned to you. I don't see how you can help. I'm sure you don't have to go home to a wife and three teenagers all wanting things that their friends have.

Choose the most appropriate response.

INTERVIEWER: You've been out of work for almost a year, and you need a job. Can you tell me what you are looking for? (Go to 13.18)

INTERVIEWER: You referred to your wife and family. How are they helping during this time? (Go to 13.19)

INTERVIEWER: You sound really frustrated, as if you had great hopes before you came here, but now you are not so sure. (Go to 13.20)

13.18 Your answer: *You've been out of work for almost a year, and you need a job. Can you tell me what you are looking for?*

Your focus on the client's problem is unlikely to succeed until you have discussed the difficulty and anger the client is displaying at being assigned to you. Return to 13.17 and try again.

13.19 Your answer: *You referred to your wife and family. How are they helping during this time?*

Although the client referred to his wife and family as additional pressures in his effort to find work, discussion of the assistance they might give him is irrelevant to the angry resistance he is displaying in the interview. Identification of the source and resolution of the resistance are required before the interview can move forward. Return to 13.17 and try again.

13.20 Your answer: *You sound really frustrated, as if you had great hopes before you came here, but now you are not so sure.*

Correct. With this response, you focus on the client's hesitance to discuss his problem with you. Before you can work together on his problem, you need to enlist his cooperation. Go to 13.21.

13.21

Clients who are uncomfortable in the interviewing situation frequently ask questions about the interview and the interviewing process. Appropriate use of information-giving responses by the interviewer can assist the client to relax and cooperate.

CLIENT: I don't see how you can help me. I've been working for 20 years as a machine operator on construction, and I don't see what you'd know about that.

INTERVIEWER: You don't think I can be of any assistance to you in finding a job.

CLIENT: You're black, young, and successful. You've probably never been out of work. What can you do to help me?

Choose the most appropriate response.

INTERVIEWER: You've identified a number of differences between us. They might interfere. We'll have to evaluate that as we go. As for whether I can help . . . at this center, we use a problem-solving approach that relies on your input and on my knowledge of the resources available to you. For the last 5 years, I have worked exclusively on placing people in construction. (Go to 13.22)

INTERVIEWER: You've identified many differences between us. You seem to be upset by the fact that I'm black, young, and successful. To be a black woman in this country doesn't guarantee success, as we all know. I've had to work very hard to get where I am. I'm sure I can help you. (Go to 13.23)

INTERVIEWER: There are a number of differences between us, but I'm not sure they are relevant to your problem. I can do little to help you if you are unwilling to help yourself. . . . Now let's see what information is needed. (Go to 13.24)

13.22 Your answer: *You've identified a number of differences between us. They might interfere. We'll have to evaluate that as we go. As for whether I can help . . . at this center, we use a problem-solving approach that relies on your input and on my knowledge of the resources available to you. For the last 5 years, I have worked exclusively on placing people in construction.*

Correct. In this response, you acknowledge and accept the legitimacy of the client's reluctance to participate and provide him with information related both to the process of assisting him to find work and to your potential to help. Go to 13.25.

13.23 Your answer: *You've identified many differences between us. You seem to be upset by the fact that I'm black, young, and successful. To be a black woman in this country doesn't guarantee success, as we all know. I've had to work very hard to get where I am. I'm sure I can help you.*

This response identifies the source of the client's reluctance to participate and provides him with information. However, the interviewer has become defensive and provided information that corresponds more with her own reaction than with the client's concerns. Return to 13.21 and try again.

13.24 Your answer: *There are a number of differences between us, but I'm not sure they are relevant to your problem. I can do little to help you if you are unwilling to help yourself. . . . Now let's see what information is needed.*

Although you have correctly identified the source of the client's discomfort and resistance, you have closed off any possibility of resolving his resistance by discounting it and blaming him for not being cooperative. By providing the client with information in response to his question, you may diffuse some of his reluctance and engage his cooperation. Return to 13.21 and try again.

13.25

As the interviewer assists the client to explore his discomfort in participating, a sense of trust develops between interviewer and client. The interviewer is then able to use more action-oriented skills, including confrontation, structuring, and reframing, to resolve resistance and enlist client cooperation. To demonstrate that resistance can be managed using a variety of skills, more than one response is correct in this frame.

CLIENT: Well it might work. . . . I'm still not sure. I've never worked with a woman before. On construction jobs, you never see women—just men. It's a strange feeling to find myself working with a woman.

Choose an appropriate response.

INTERVIEWER: It's good to know that, after nearly a year out of work, you still have the energy and motivation to confront problems—like working with a woman. That energy will be useful in finding an appropriate job. (Go to 13.26)

INTERVIEWER: You find yourself in a difficult situation. On the one hand, you need assistance to find a job, and on the other hand, you are not completely comfortable working with a woman. Deciding whether to work with me is not easy. (Go to 13.27)

INTERVIEWER: New experiences like working with a woman can be very anxiety provoking, particularly when it is so important for you to find a job. (Go to 13.28)

13.26 Your answer: *It's good to know that, after nearly a year out of work, you still have the energy and motivation to confront problems—like working with a woman. That energy will be useful in finding an appropriate job.*

Correct. Resistance can be seen as a sign of life, and presenting the client with this alternative perspective of his resistance can enable him to utilize it in the achievement of his goals.

The other two interviewer responses in this frame present alternative ways of working toward client cooperation. You may wish to review them before moving on. Go to 13.29.

13.27 Your answer: *You find yourself in a difficult situation. On the one hand, you need assistance to find a job, and on the other hand, you are not completely comfortable working with a woman. Deciding whether to work with me is not easy.*

Correct. This combined reflection and confrontation focuses the client's attention on his current behavior and encourages him to come to a resolution of his resistance.

The other two interviewer responses in this frame present alternative ways of working toward client cooperation. You may wish to review them before moving on. Go to 13.29.

13.28 Your answer: *New experiences like working with a woman can be very anxiety provoking, particularly when it is so important for you to find a job.*

Correct. This summarization, which focuses on both the client's resistance and his problem, will help him decide whether he wants to spend more time on his discomfort in working with a woman or move on to work on his unemployment problem.

The other two interviewer responses in this frame present alternative ways of working toward client cooperation. You may wish to review them before moving on. Go to 13.29.

13.29

So far in this chapter, you have learned to identify client reluctance and to use single skills and combinations of skills in enlisting client cooperation. Additionally, you can use the general interviewing strategy—structuring for information and action—to assist clients at those phases of the interview in which they experience difficulties with cooperation. When reluctance to participate is high and likely to prevent any successful outcome, the skill of structuring can be particularly important because it enables clients to participate in the attainment of each action objective associated with the resolution of their resistance.

To help illustrate the application of structuring for information and action to resistance, Table 13.2 contains an outline of the example to be used in the remaining frames of this chapter. These frames center on a 32-year-old parent who has been found guilty of physically abusing a 5-year-old child. The parent has been referred by the court to meet with a child-care worker on a regular basis. As shown in Table 13.2, the client has more than one area of resistance to be resolved. The frames that follow focus on only one of these concerns.

During the first interview, interviewer and client identified the client's reluctance to participate. They agreed that this reluctance must be considered before the client's abusive behavior can be addressed. They are now reaching the conclusion of the development phase.

CLIENT: No matter how hard I try, I just can't get over my fear of the law and what it can do to me and my family.

Choose the most appropriate structuring response.

INTERVIEWER: Now that we know what your concern is, I think we can get back to the problems you have with managing your children. Is that agreeable to you? (Go to 13.30)

INTERVIEWER: You've identified that a major problem in meeting with me is your fear of the law. Let's talk about this in more detail and specifically identify what you are so fearful about. Are you open to that? (Go to 13.31)

INTERVIEWER: You've identified that a major problem in meeting with me is your fear of the law. Perhaps you can suggest ways in which we can get around this problem. Are you open to that? (Go to 13.32)

13.30 Your answer: *Now that we know what your concern is, I think we can get back to the problems you have with managing your children. Is that agreeable to you?*

Unless you work through the stages of structuring to the point at which you and the client implement an action that successfully manages the client's reluctance to participate, cooperation in learning new ways to manage children will not occur. In this response, you return to the child-management issue prematurely, before the client's reluctance to participate has been resolved. Return to 13.29 and try again.

13.31 Your answer: *You've identified that a major problem in meeting with me is your fear of the law. Let's talk about this in more detail and specifically identify what you are so fearful about. Are you open to that?*

Correct. This structuring response invites the client to develop a more specific definition of the problem that results in difficulty with participating. Go to 13.33.

Table 13.2

Example Outcomes at Each Interviewing Phase When Structuring for Information and Action Is Applied to Resistance

Interviewing phase	Example outcome
Exploratory	Client states a fear of the court system Client indicates shame at what has happened
Problem Identification	Client identifies the fact that she or he does not want to lose children Client concludes that what has happened is against a personal value that children should not be abused
Priorization of Problems	Concern about loss of children because information is given to the court by interviewer Rationalization of values with what has occurred
Goal-formulation	To retain custody of children and participate in the interviewing process To endorse values and to use them to prevent future acts of abuse
Preparation for Action	To renegotiate client–interviewer contract with the court to provide greater confidentiality To focus on client's values about child abuse later in the interviewing process when child management is discussed
Action	To petition court to have a child-care worker monitor abuse and an independent counselor develop child-management skills with the client

---

13.32 Your answer: *You've identified that a major problem in meeting with me is your fear of the law. Perhaps you can suggest ways in which we can get around this problem. Are you open to that?*

This response invites the client's participation but moves too quickly to the action phase of problem resolution. Before formulating an action plan, client and interviewer need to develop a more specific definition of the problem and to identify an attainable goal associated with it. Return to 13.29 and try again.

13.33

In the inventory phase, the client has identified a specific concern—that custody of the children may be lost because the child-care worker is required to regularly report to the court. Resolution of this concern appears to be the immediate priority.

CLIENT: I agree. I'm not willing to talk about my children or myself as long as you have to report it all to the court.

Choose the most appropriate structuring response.

INTERVIEWER: Now that we have agreed on the major difficulty you are having, it's important for us to restate it as a goal. What do you think is a reasonable goal toward which we can work? (Go to 13.34)

INTERVIEWER: Now that we have agreed about your difficulty, I think it's important that you understand that neither of us has any choice. (Go to 13.35)

INTERVIEWER: Now that we are aware of what the problem is, you should know that, as long as we meet regularly and you make an effort, I can help by submitting good reports. Does that seem reasonable? (Go to 13.36)

13.34 Your answer: *Now that we have agreed on the major difficulty you are having, it's important for us to restate it as a goal. What do you think is a reasonable goal toward which we can work?*

Correct. This structuring response moves the interview from the priority to the goal-formulation phase of the interviewing process. Achieving an agreed-on goal that you and the client can work toward requires the effective use of your interviewing skills throughout the phase. The "goal-formulation" section in Table 13.2 states possible goals. Go to 13.37.

13.35 Your answer: *Now that we have agreed about your difficulty, I think it's important that you understand that neither of us has any choice.*

This response is an incorrect statement of fact. All clients, even involuntary clients, have a choice as to whether they participate in therapy, even though they may not like the outcome of their choice. It is important that interviewers assist clients in formulating goals to confront and resolve resistance when the clients feel as though they are under legal, moral, or psychological coercion. Return to 13.33 and try again.

13.36 Your answer: *Now that we are aware of what the problem is, you should know that, as long as we meet regularly and you make an effort, I can help by submitting good reports. Does that seem reasonable?*

By using this response, you attempt to coerce and placate the client into participating in the interview. This does not assist the client in formulating a goal related to resolution of the resistance. Return to 13.33 and try again.

13.37

During the goal-formulation phase, the client and the interviewer, after some discussion, have agreed that a reasonable goal for the client is to both retain custody of the children and participate in the interviewing process. Client and interviewer have discussed the merits of a number of action plans, and a suitable action plan is emerging.

CLIENT: I like working with you, but if you must report to the court all the time, then I prefer to talk to someone else about my day-to-day problems.

Choose the most appropriate structuring response.

INTERVIEWER: We seem to agree that I should check with the court about whether you can work with another counselor on child management while I report on the child-abuse issue. I am upset and a little perplexed that you don't feel able to work with me on both issues. (Go to 13.38)

INTERVIEWER: We seem to agree that I should check with the court about whether you can work with another counselor on child management while I report on the child-abuse issue. You should be aware that not all child-care workers would agree to do this. (Go to 13.39)

INTERVIEWER: We seem to agree that I should check with the court about whether you can work with another counselor on child management while I report on the child-abuse issue. If it's agreeable to you, I'll make an application to the court. (Go to 13.40)

13.38 Your answer: *We seem to agree that I should check with the court about whether you can work with another counselor on child management while I report on the child-abuse issue. I am upset and a little perplexed that you don't feel able to work with me on both issues.* By making the last statement, you indicate an inability to manage your own feelings while the issue of client resistance is being resolved. Rather than facilitating client cooperation, you pressure the client, which can lead to heightened resistance. Return to 13.37 and try again.

13.39 Your answer: *We seem to agree that I should check with the court about whether you can work with another counselor on child management while I report on the child-abuse issue. You should be aware that not all child-care workers would agree to do this.*

The initial part of your response summarized the plan that you and the client formulated to help resolve the client's reluctance to participate. Your closing statement, however, reflects your difficulty and discomfort with the plan. Unless you confront and resolve your discomfort, it is likely to inhibit the client's progress. Return to 13.37 and try again.

13.40 Your answer: *We seem to agree that I should check with the court about whether you can work with another counselor on child management while I report on the child-abuse issue. If it's agreeable to you, I'll make an application to the court.*

Correct. In this structuring response, you summarize the conclusion reached in the preparation for action phase and invite the client to enact the plan. Go to 13.41.

13.41

When trust and understanding have been experienced by the client, it is not unusual for the client to display discomfort and reluctance during the termination phase. The effective interviewer understands and supports the client during the resolution of this discomfort.

The child-care worker has informed the client that the court has decided to accept the proposed plan of separate workers for the issues of child management and child abuse. The interview arranged to discuss this decision is drawing to a close.

CLIENT: I'm not sure that the arrangements we made are the best after all. I'm growing used to the idea of the court knowing everything. . . . Maybe it wouldn't matter. I've also grown used to you. How am I ever going to get to know another worker?

Choose the most appropriate response.

INTERVIEWER: You've grown to trust me even though I represent the court to you. Maybe we can arrange matters so that you can work with me. (Go to 13.42)

INTERVIEWER: You've grown to trust me. You're wondering whether coping with our discussions being open to the court is better than having to learn to trust yet another person. (Go to 13.43)

INTERVIEWER: You seem to be changing your mind a great deal. It's not possible for me to keep changing things with the court. Exactly what is it that you want? (Go to 13.44)

13.42   Your answer: *You've grown to trust me even though I represent the court to you. Maybe we can arrange matters so that you can work with me.*

Although this outcome may enhance your self-esteem for the moment, in the long term the client's resistance, which is related to your dual responsibility, will resurface. The agreed-on plan is a reasonable one and should be tried before other alternatives are considered. You need to confront the client's resistance to termination with you and work to facilitate cooperation with the new interviewer. Return to 13.41 and try again.

13.43   Your answer: *You've grown to trust me. You're wondering whether coping with our discussions being open to the court is better than having to learn to trust yet another person.*

Correct. This summarization of the client's concerns over termination will assist in resolving the discomfort of leaving the interviewing relationship established with you. With the resolution of the resistance associated with confidentiality, the client is able to move forward with the new interviewer to talk about the difficulties of managing children. Go to 13.45.

13.44   Your answer: *You seem to be changing your mind a great deal. It's not possible for me to keep changing things with the court. Exactly what is it that you want?*

The client has expressed discomfort over termination, which is not unusual when an interviewer has been helpful. Your expression of frustration does not assist in resolving reluctance. In this concluding phase of the interviewing process, the client requires understanding and support so that a comfortable transition can be made to another child-care worker. Return to 13.41 and try again.

13.45

All clients respond to the interviewing process with some degree of reluctance or discomfort. This is normal. Participation in an interview can be stressful, whether it is voluntary or involuntary, sought or imposed, desired or required. Even highly motivated clients may be resistant to discussion of certain topics. The effective interviewer accepts this, uses the skills learned in previous chapters, and assists the client to work through emotions associated with the discomfort without giving premature advice or false reassurance.

Recently, it has been shown that there are five stages that individuals progress through when making changes in their lives. Individuals may come to the interview at any one of these stages. Recognition of the stage at which a client is functioning will enable the interviewer to enlist the client's cooperation more effectively. The stages of change, the challenges to cooperation at each stage, and the goals and methods found useful at each stage are shown in Table 13.3.

The next sequence of frames is designed to illustrate the use of enlisting cooperation to assist a client through each of the stages of change. The following frames focus on a series of interviews between a psychologist and a senior manager in a financial institution. The manager has been referred because of a recent heart attack and a request by a cardiologist for a change in lifestyle.

It is not unusual that clients come to an initial interview at the precontemplation stage. Accordingly, during the first interview the goal is often to assist the client to acknowledge there is a problem.

CLIENT: (glaring, and with a raised voice) I don't know why the cardiologist has sent me to see you. I have to get back to work before things start to go wrong again. I don't have time to waste on visiting a psychologist.

Choose the most appropriate response.

INTERVIEWER: I understand how important it is for you to return to work, but it's not good for you to be so involved in your job. You need other interests in your life. (Go to 13.46)

INTERVIEWER: Don't worry about why you are here. I'm sure that if you just go along with the referral, you'll be surprised at the outcome. (Go to 13.47)

INTERVIEWER: It sounds as if you being here is an imposition, and you're rather angry about it. Perhaps it would help if we could talk about the reason you were referred. (Go to 13.48)

13.46 Your answer: *I understand how important it is for you to return to work, but it's not good for you to be so involved in your job. You need other interests in your life.*

This response fails to recognize the client's emotional response and reluctance to participate. The interviewer has given advice based on assumptions about the client's lifestyle, which may or may not be true. The client is likely to respond with even more emotion and reluctance, failing to move beyond the precontemplation stage. Return to 13.45 and try again.

13.47 Your answer: *Don't worry about why you are here. I'm sure that if you just go along with the referral, you'll be surprised at the outcome.*

Although your experience may support your statement, it will do little to convince the client that seeing you will have any benefit. You need to help the client understand the reason for the referral. Return to 13.45 and try again.

Table 13.3

Stages of Change, Challenges to Cooperation, and Goals and Methods

Stage	Challenges to cooperation	Goal and method
Precontemplation: Clients are unaware they have a problem, are aware they have a problem but do not intend to do anything about it, or feel it can be solved by others changing their behavior.	Clients do not believe they have a problem. Clients attend the interview as a result of pressure by others or agencies to seek help.	Assist the client to own the problem by encouragement and the discussion of a rationale for change and the positive benefits of change.
Contemplation: Clients are beginning to acknowledge that they have a problem, but are not yet committed to doing anything about it. No method of dealing with the problem has been developed.	Clients do not see change as imminent. They see the need for change, but may not believe they need to be part of the solution. Clients often debate the benefits versus the cost of change.	Assist clients to move from discussion of the problem toward preparing to act on the problem by educating the client about the change process and enhancing the benefits of change while challenging the disadvantages.
Preparation for Action: Clients have decided to work toward a goal and are actively developing an action plan.	Clients have difficulty formulating a workable action plan or moving from planning to action.	Assist the client to develop a workable action play by ensuring that all alternative action plans are considered and evaluated.
Action: Clients execute the action plan.	Clients fail to effect the action plan either because they experience it as too simple or too complex, or are unable to modify the steps involved.	Assist clients to initiate the action plan, monitor it, and modify the action plan as required using the skills learned.
Maintenance: Clients actively work to maintain the change made during the action phase.	Clients are unable to maintain the changes they have made during the action phase.	Assist clients to establish a program through which they can maintain the changes that have been made.

13.48 Your answer: *It sounds as if you view being here as an imposition, and you're rather angry about it. Perhaps it would help if we could talk about the reason you were referred.*

Correct. By using reflections of content and feeling, you have both acknowledged the client's reluctance to participate and provided support for exploration and management of the discomfort associated with the referral. Your response also provides the client with an opportunity to consider the possibility that there is a reason for the referral. This will help the client to move through the precontemplation stage. Go to 13.49.

13.49

When a client is reluctant or uncooperative, the interviewer may experience unexpected feelings, including confusion, panic, hostility, and rejection. Management of these feelings and focus on the client's resistance are important to a successful outcome at any stage in the interviewing process, but particularly during the precontemplation stage.

CLIENT: Of course I'm angry. Here I am wasting time with you. I've had a heart attack, not a head injury. What reason is there to meet with you?

Choose the most appropriate response.

INTERVIEWER: Your cardiologist made the referral. If you're not happy being here, you should make other arrangements. (Go to 13.50)

INTERVIEWER: You're upset about being referred, but in answer to your question, people with cardiac problems can often improve their health by examining and changing their lifestyles. (Go to 13.51)

INTERVIEWER: I find it frustrating to work with cardiac clients. You are never cooperative, and you all think you'll be fine. (Go to 13.52)

13.50 Your answer: *Your cardiologist made the referral. If you're not happy being here, you should make other arrangements.*

When managing resistance in the precontemplation stage, the interviewer is aware that the client's responses may arouse unexpected feelings. The interviewer continues to focus on the client and guards against the expression of unhelpful statements that could further increase the client's resistance. Return to 13.49 and try again.

13.51 Your answer: *You're upset about being referred, but in answer to your question, people with cardiac problems can often improve their health by examining and changing their lifestyles.*

Correct. You have provided support and encouraged the client to think about the benefits of the referral. By providing information you encourage the client to move from the precontemplation stage to the contemplation stage. Go to 13.53.

13.52 Your answer: *I find it frustrating to work with cardiac clients. You are never cooperative, and you all think you'll be fine.*

This inappropriate expression of feeling on your part may help calm you, but it depersonalizes the client and will do little to assist in the resolution of the client's resistance. Return to 13.49 and try again.

13.53

Lack of cooperation at a particular phase of interviewing does not indicate that the client is unmotivated. Effective interviewers recognize that resistant behaviors can provide self-protection and day-to-day life continuity for clients. Interviewers must not seek to blame clients but rather use their skills to assist clients to work on issues. In this way, a climate of cooperation can be established, and clients are enabled to move from stage to stage.

CLIENT: I guess I sound angry, but I'm not angry at you. I just don't know how I'm going to manage all my responsibilities at work. I was in the hospital for 3 weeks. I've been told I can only work part time, and now I'm supposed to see you. I suppose there are changes to be made, but I'm not sure I can do much until other things are resolved.

Choose the most appropriate response.

INTERVIEWER: Your illness has resulted in lots of sudden change in your life. The thought of possibly more change feels overwhelming. . . . It's all too much. (Go to 13.54)

INTERVIEWER: Despite your illness and the difficulties it has caused you, you seem unwilling to make any changes that might prevent future problems. (Go to 13.55)

INTERVIEWER: It sounds as if, no matter how hard I try, we're not going to achieve much because you are too concerned about problems at work. (Go to 13.56)

13.54 Your answer: *Your illness has resulted in lots of sudden change in your life. The thought of possibly more change feels overwhelming. . . . It's all too much.*

Correct. This summary statement focuses the interview on the source of the client's resistance to participating. It conveys your understanding of the client's difficulty in cooperating and will facilitate discussion on difficulties in changing, a step toward the contemplation stage. Go to 13.57.

13.55 Your answer: *Despite your illness and the difficulties it has caused you, you seem unwilling to make any changes that might prevent future problems.*

A challenge of this nature may be appropriate at a later time; however, early in the interview it provides little support or encouragement to a client struggling with resistance to change. Return to 13.53 and try again.

13.56 Your answer: *It sounds as if, no matter how hard I try, we're not going to achieve much because you are too concerned about problems at work.*

The client has experienced many sudden life changes and is expressing discomfort and upset over the problems that these and potential future changes might bring. Blaming the client for lack of progress in the interviewing process will cause alienation of the client and no movement toward the contemplation stage. Return to 13.53 and try again.

13.57

During the contemplation stage, clients being to acknowledge that they have problems but show ambivalence or little willingness to commit to change. To enlist client cooperation during this stage the interviewer uses the skills already learned to encourage clients to consider the advantages of making changes in their lives. When clients have difficulty in preparing to act on a problem, action-oriented skills such as confrontation and information giving can be used to challenge their ambivalence.

Interviewer and client have now met on two separate occasions. During this time, the client has discussed both the reasons for referral and the hectic lifestyle.

CLIENT: Since coming to see you I've realized that I need to change the way I live, including many of the commitments I have at work, but I can't do anything until I get back to work full time. They are asking me all the time when I will be back full time.

Choose the most appropriate response.

INTERVIEWER: That's a very good idea. I am sure you will feel much less stressed as soon as you have cleared all the work that has accumulated on your desk. (Go to 13.58)

INTERVIEWER: I am a little confused. On the one hand you state that you have become aware of the necessity of making change in your life, but on the other hand you state that you wish to return to full time work as quickly as possible. Can you help me to understand this? (Go to 13.59)

INTERVIEWER: I find that when I need to make change in my life, it is important to clear my desk and find some time to think through the changes. (Go to 13.60)

13.58 Your answer: *That's a very good idea. I am sure you will feel much less stressed as soon as you have cleared all the work that has accumulated on your desk.*
Your response has reinforced the client behavior that contributed to the cardiac problems. Your task is to assist the client to commit to change and prepare to establish new behaviors that will lead to a healthier lifestyle. Return to 13.57 and try again.

13.59 Your answer: *I am a little confused. On the one hand you state that you have become aware of the necessity of making change in your life, but on the other hand you state that you wish to return to full-time work as quickly as possible. Can you help me to understand this?*
Correct. The client has now acknowledged that there is need for change but has shown a lack of commitment to do so. By identifying the discrepancy in the client's goals you have challenged the client to consider the ambivalence inherent in the response, namely the stated need to change and the decision to delay action in changing priorities. This will assist the client to focus on the more immediate action required. Go to 13.61.

13.60 Your answer: *I find that when I need to make changes in my life, it is important to clear my desk and find some time to think through the changes.*
This self-disclosure, which endorses the client's position, will do little to focus on the needed change in priorities, including the need to consider the importance

of the many current commitments at work. At this point it is important to employ the skills you have learned to assist the client to focus on a change in priorities and lifestyle in the near future. Return to 13.57 and try again.

13.61

In the preparation for action stage, clients decide to work toward a goal and actively plan what action to take. Some clients may have difficulty forming a workable action plan, whereas others have difficulty in moving forward with the plan.

CLIENT: As I think about it, returning to full-time work is asking for a disaster. Perhaps it is more important to reorganize my priorities and to consider the importance of some of the things I am currently engaged in at work. I have no idea where to start.

Choose the most appropriate response.

INTERVIEWER: How about starting with all of those things that you don't like doing or that create the most stress for you. (Go to 13.62)

INTERVIEWER: It can be overwhelming and anxiety provoking when you decide to do something but have no idea where to start. Perhaps we should talk a little bit about your anxiety. (Go to 13.63)

INTERVIEWER: Well, what many find effective is to generate a number of ways to attain a goal, evaluate them, and then select the strategy that is likely to work the best. I wonder if this might work for you. (Go to 13.64)

13.62 Your answer: *How about starting with all of those things that you don't like doing or that create the most stress for you.*

The client will be best assisted if enabled to develop a workable plan for a reorganized life, rather than being provided with ideas that have not been evaluated. Teaching clients to prepare for action is an important skill. Return to 13.61 and try again.

13.63 Your answer: *It can be overwhelming and anxiety provoking when you decide to do something but have no idea where to start. Perhaps we should talk a little bit about your anxiety.*

This reflection of feeling would be appropriate if the client had shown any anxiety about proceeding to develop an action plan. The client has expressed an interest in proceeding, but a lack of knowledge about how to proceed. Return to 13.61 and try again.

13.64 Your answer: *Well, what many find effective is to generate a number of ways to attain a goal, evaluate them, and then select the strategy that is likely to work the best. I wonder if this might work for you.*

Correct. This information-giving response provides the client with the information necessary to establish a plan of action. Together client and interviewer can generate, select, and evaluate the strategies for the client to implement in the action stage. Go to 13.65.

13.65

During the action stage, clients act to attain their goal. Many encounter success, whereas others require emotional support and assistance to modify the plan. Continued use of the skills you have learned will enable you to assist clients in this stage.

The interviewer and client have discussed a variety of action plans and after a careful evaluation of each, the client has decided on the following actions:

- *not to return to full-time work at this time*
- *when at work, to delegate tasks to colleagues*
- *to join an exercise program for cardiac patients*
- *to be less controlling*
- *to establish some relaxing leisure-time activities*

It is now a week later and the interviewer and client are reviewing progress with the implementation of the action plan.

CLIENT: I tried all week to do the things I decided on at our last session, but the more I thought about it there were just too many things to do. I assigned a few tasks to others at work, but that is all. I am not very happy with my progress.

Choose the most appropriate response.

INTERVIEWER: You tried very hard to implement the plan. You had some success at work, but found it too hard to implement all of the tasks you had established for yourself. These are large goals you have set. What do you think about establishing some subgoals or priorities? (Go to 13.66)

INTERVIEWER: You seem quite unhappy with your progress toward your goal. Have you always been hard on yourself when you fail to meet your goals? (Go to 13.67)

INTERVIEWER: It is very pleasing to hear that you tried so hard and achieved success at work. It is important that you keep on trying. I am sure that by next week you will have achieved success with your other goals. (Go to 13.68)

13.66 Your answer: *You tried very hard to implement the plan. You had some success at work, but found it too hard to implement all of the tasks you had established for yourself. These are large goals you have set. What do you think about establishing some subgoals or priorities?*

Correct. In this response you used reflections of feeling and content and an open question in assisting the client to consider a way to modify the action plan. Your response will enable the client to establish a set of steps to accomplish the overall goal and to enhance cooperation in achieving change. Go to 13.69.

13.67 Your answer: *You seem quite unhappy with your progress toward your goal. Have you always been hard on yourself when you fail to meet your goals?*

It is unlikely that concentrating on the client's feelings at this point will be productive. The closed question about past emotional responses to failure is unlikely to assist the client to revise the action plan and achieve success. Return to 13.65 and try again.

13.68 Your answer: *It is very pleasing to hear that you tried so hard and achieved success at work. It is important that you keep on trying. I am sure that by next week you will have achieved success with your other goals.*

The client has informed you that difficulties were experienced in attaining the goals that were agreed upon and that lack of success resulted in unhappiness. Your task at this stage is to assist the client to modify the steps involved in the action plan in order to achieve success.

13.69

Although it is often difficult for clients to make changes in their lives, the rewards of attaining change can be very satisfying. Perhaps more difficult for some clients is the maintenance of the changes they have made. Some clients experience no difficulty maintaining change, whereas others will require your assistance in developing strategies through which they can maintain the changes made.

After several sessions, the client has established the changes made in the action stage. The client continues to work part time, has begun to delegate tasks to others, is exercising regularly, and is involved in some relaxing activities. On occasion the client has difficulty assigning new tasks to others, preferring to retain full control of those tasks.

CLIENT: I am glad that I have been able to make these changes in my life; however, I still struggle with the assignment of tasks to others. Often the new tasks are challenging and I am reluctant to pass them on to others, who might not be as effective as I am, or who might jeopardize the careers of those involved.

Choose the most appropriate response.

INTERVIEWER: You've made good progress in attaining your goals. Perhaps we can terminate our interviews now and review your situation in 3 months. At that point you will probably have attained all of your goals successfully. (Go to 13.70)

INTERVIEWER: It is good to hear that you have established and maintained most of the new behaviors for which you planned. Giving up rewarding tasks to others appears to be difficult and continues to be a constant struggle for you. What do you think might make you more comfortable in giving up the responsibility for new tasks to others? (Go to 13.71)

INTERVIEWER: You have made good progress in attaining most of your goals. However, I am wondering what makes you think that you are more effective than others in your workplace. (Go to 13.72)

13.70 Your answer: *You've made good progress in attaining your goals. Perhaps we can terminate our interviews now and review your situation in 3 months. At that point you will probably have attained all of your goals successfully.*

This structuring response suggests that you have decided to ignore the difficulty the client is having with an important area of change. Rather than ignore this area, it is important that you help the client establish a program to maintain the assignment of new tasks to others. Return to 13.69 and try again.

13.71 Your answer: *It is good to hear that you have established and maintained most of the new behaviors for which you planned. Giving up rewarding tasks to others appears to be difficult and continues to be a constant struggle for you. What do you think might make you more comfortable in giving up the responsibility for new tasks to others?*

Correct. Your reflection of feeling and content support the changes the client has made. The open question invites discussion of the one aspect of change with which the client is struggling. Fostering the maintenance of the changes the client has implemented is important at this stage of the change process. The client and interviewer now proceed to set up a program to maintain the client's ability to delegate tasks. Go to 13.73.

13.72 Your answer: *You have made good progress in attaining most of your goals. However, I am wondering what makes you think that you are more effective than others in your workplace.*

This response raises an interesting direction for questioning, but as phrased could be viewed negatively by the client. When clients feel devalued by the interviewer they are less likely to cooperate with the development of strategies to maintain new behaviors, the task at this stage. Return to 13.69 and try again.

13.73

Everyone experiences some reluctance, resistance, and discomfort at being interviewed. During the course of our lives, we have all developed biases and values that can result in instinctive resistances. The effective interviewer is aware of possible sources of these resistances and works supportively and creatively to resolve them. Highly resistant clients present a particular challenge to interviewers. Effective interviewers are careful not to argue with clients and make every effort to turn resistance to advantage. Although use of the skills taught in this chapter will assist interviewers with many resistant clients, highly resistant clients may require the assistance of advanced interviewing and/or intervention techniques. Go to 13.74.

13.74

Review Questions

Indicate whether each of the following statements is true or false.

1. Resistant behaviors may provide self-protection for clients.
2. Appropriate use of information-giving responses can enlist the cooperation of resistant clients who ask questions.
3. Clients who provide long responses are never resistant.
4. When interviewers experience hostility, expression of this feeling is important to a successful outcome.
5. Some degree of client discomfort or reluctance to an interview is normal.
6. Structuring for information and action cannot be used to enlist client cooperation.
7. When a sense of trust has been developed, action-oriented skills can be used to resolve resistance and enlist client cooperation.

8. By the manner in which they communicate, clients may demonstrate their reluctance to cooperate.
9. During the action stage many clients require support to modify their action plan.

Review Answers

1. *True.* If you answered *false*, go to 13.53 and review.
2. *True.* If you answered *false*, go to 13.21 and review.
3. *False.* If you answered *true*, go to 13.1 and review.
4. *False.* If you answered *true*, go to 13.49 and review.
5. *True.* If you answered *false*, go to 13.17 and review.
6. *False.* If you answered *true*, go to 13.29 and review.
7. *True.* If you answered *false*, go to 13.25 and review.
8. *True.* If you answered *false*, go to 13.9 and review.
9. *False.* If you answered *true*, go t 13.65 and review.

If six or more of your responses were incorrect, you should return to 13.1 and review the material in this chapter.

Points to Remember about Enlisting Cooperation

1. WHEN IDENTIFYING LACK OF CLIENT COOPERATION:
   a. Focus on the amount of information given by the client.
   b. Focus on whether the client's discussion is meaningful or restricted.
   c. Focus on the way in which the client interacts.
   d. Focus on whether the client keeps or violates the formal or informal interviewing contract.
   e. Focus on the client's stage of change.
2. WHEN ENLISTING COOPERATION:
   a. Anticipate some degree of reluctance from all clients.
   b. Monitor and manage your own feelings.
   c. Do not confuse resistance with lack of motivation.
   d. Identify the client's stage of change and respond accordingly.
3. TO ENLIST COOPERATION:
   a. Focus on the issue of resistance before the purpose of the interview.
   b. Use a wide range of basic interviewing skills and action-oriented skills.
   c. Use structuring for information and action to resolve resistance with difficult clients.
   d. Facilitate the client's movement through the stages of change.

13.1

Activity Unit

Find two others who have also read this chapter. Role play short interviews alternating the client, interviewer, and observer roles. Record each brief interview on audio- or videotape. An interview with at least five interviewer responses is sufficient for this exercise, although more responses may be required if the role played is that of a highly resistant client.

Interviewer role　Practice enlisting cooperation with a resistant client.

Client role　Be uncooperative (but not too uncooperative) so that the interviewer can practice using interviewing skills to enlist your cooperation. Clients may want to role play the problem of a friend or family member, using any of the sources of client resistance outlined in Table 13.1, or to enact one of the following roles:

1. An adolescent, on probation for shoplifting, meeting with a probation officer.
2. A pregnant alcoholic who has been referred by her physician because of her drinking problem.
3. An abused wife who is unable to talk about her husband.
4. An employee who is referred to the personnel office because his or her current job is to be phased out.

Observer role　Operate the equipment and complete the Practice Interview Checklist (see Table 13.4).

Postinterview discussion　Using the taped interview and the Practice Interview Checklist data, the participants should discuss the performance of the interviewer. All participants can learn from this discussion, which should be nonjudgmental, focus on the positive as well as less effective responses, and stimulate improved performance for all.

13.2

Activity Unit

Using the tape of the interview produced in Activity Unit 13.1 or the tape of another practice interview, enter the interview responses used to enlist client cooperation. Then formulate a better response to replace each original one in the space provided.

Original response 1 _____

_____

_____

Better response _____

_____

_____

Original response 2 _____

_____

_____

Table 13.4

Practice Interview Checklist

Activity	Interviewer response[a]				
	1	2	3	4	5
Appropriate nonverbal behavior					
On topic					
Response enlisting cooperation					
Effective response[b]					

[a] Check the activities that characterize each response.

[b] Indicate the degree of effectiveness on a scale from 1 to 5 in which 1 is ineffective and 5 is very effective.

Better response _____

_____

_____

Original response 3 _____

_____

_____

Better response _____

_____

_____

Original response 4 _____

_____

_____

Better response _____

_____

_____

Original response 5 _____

_____

_____

Better response _____

_____

_____

Cultural Considerations

1. Be aware that reasons for resistance differ from culture to culture and with a client's level of acculturation.
2. When enlisting cooperation, it is important to discuss cultural and acculturation issues.

InfoTrac® College Edition Keyword Search Terms

Denial AND Counseling
Resistance AND Counseling

Chapter Fourteen

Putting It All Together

The preceding chapters deal with a number of skills that are essential to successful interviewing: attending behavior, effective questioning, reflecting content and feeling, communicating feeling and immediacy, confronting, self-disclosing, information giving, interpreting, structuring, and enlisting cooperation. The material in this chapter is intended to help you integrate these skills.

After completing this chapter, you should be able to:

1. Use the skills presented in this book in a systematic and integrated manner.
2. Decide when it is appropriate to use a given skill.

The program in this chapter focuses on an interview between an unemployed man and a counselor in a crisis center. The client has been unemployed since he and his family moved 6 months ago, and he now has a severe financial problem. This is the first time he has discussed his problem with anyone.

Because there are several routes through this chapter, more than one response may be correct in each frame. Go to 14.1.

14.1   INTERVIEWER: How can I help you today?

CLIENT: I'm not sure you can help me, but I've got to discuss this with someone. . . . I just can't find a job. I've been looking for 6 months now. I look through the want ads every day, but there's just nothing. We have three children, my wife isn't able to work right now, and our savings are just about gone.

Choose an appropriate response.

INTERVIEWER: Have you gone to the unemployment office regularly? (Go to 14.2)

INTERVIEWER: You sound desperate. Could you help me to understand your situation better? (Go to 14.3)

INTERVIEWER: Your financial situation is grim, and you're wondering what to do next. (Go to 14.4)

14.2   Your answer: *Have you gone to the unemployment office regularly?*
This closed question diverts the client from an exploration of his problem and demonstrates poor attending behavior. At this point in the interview, you should encourage the client to explore his problem with you. Return to 14.1 and try again.

14.3   Your answer: *You sound desperate. Could you help me to understand your situation better?*
Correct. This reflection of feeling, followed by an open question, enables the client to feel comfortable while discussing his problem with you. Go to 14.6.

14.4   Your answer: *Your financial situation is grim, and you're wondering what to do next.*
Correct. This restatement indicates that you are following the client and invites him to continue discussing his problem with you. Go to 14.5.

14.5   CLIENT: It's really crucial that I get a job. I'm sure there must be some way for me to get back into the real estate business. There must be something I can do.

Choose an appropriate response.

INTERVIEWER: It sounds as though you're under a great deal of pressure and that getting a job is of major importance to you. That kind of pressure often creates other difficulties. (Go to 14.7)

INTERVIEWER: Your situation sounds really stressful. (Go to 14.8)

INTERVIEWER: You could take care of the children while your wife goes out to work. Have you thought about that? (Go to 14.9)

14.6   CLIENT: I'm getting desperate. I lost my job about 6 months ago, so I thought I'd come here and get back into real estate. I renewed my license and went around to all the companies, but there just aren't any openings. I have to find work soon. I don't have any money left.

Choose an appropriate response.

INTERVIEWER: Have you thought about setting up your own business? That might be a reasonable way to solve your problem. (Go to 14.10)

INTERVIEWER: So, you've tried to make use of your skills. (Go to 14.11)

INTERVIEWER: You tried to improve your situation by moving, but things have grown worse. (Go to 14.12)

14.7   **Your answer:**   *It sounds as though you're under a great deal of pressure and that getting a job is of major importance to you. That kind of pressure often creates other difficulties.*
Correct. This reflection of feeling and content indicates that you are following the client and are interested in what he is telling you. Also, you've invited him to explore other difficulties. Go to 14.15.

14.8   **Your answer:**   *Your situation sounds really stressful.*
Correct. This reflection of feeling indicates that you are listening to the client and encourages him to explore his situation further. Go to 14.16.

14.9   **Your answer:**   *You could take care of the children while your wife goes out to work. Have you thought about that?*
This response shows that you haven't been listening to the client: he has told you that his wife isn't able to work. Return to 14.5 and try again.

14.10   **Your answer:**   *Have you thought about setting up your own business? That might be a reasonable way to solve your problem.*
This may be a reasonable solution, but at this early stage of the interview, you need more information about the client. Return to 14.6 and try again.

14.11   **Your answer:**   *So, you've tried to make use of your skills.*
Correct. Your brief restatement indicates that you are attending to the client and invites him to continue. Go to 14.13.

14.12   **Your answer:**   *You tried to improve your situation by moving, but things have grown worse.*
Correct. This reflection of content shows the client that you have heard what he has said and encourages him to discuss his difficulties further. Go to 14.14.

14.13

CLIENT: I really don't want to go back to working in a factory. I want a clean job where noisy machinery isn't going all the time, and one in which I can meet and mix with people. But I'm so desperate, I'd take anything at the moment. I guess I've got two big problems right now: I can't find a job, and I don't have any money.

Choose an appropriate response.

INTERVIEWER: So, you have two urgent problems. Could you help me to understand which one is more pressing? (Go to 14.17)

INTERVIEWER: Everyone seems to be in the same situation these days. Is there anyone you could borrow money from? (Go to 14.18)

INTERVIEWER: I get the feeling that you really haven't been trying hard enough to find a job. What things did you say you'd done so far? (Go to 14.19)

14.14

CLIENT: Yes. We're a lot worse off. We're renting a small house that's cold and damp, and my wife and I can't go out because we can't afford a sitter. But these are small problems compared to being broke and out of work. I guess my biggest problem right now is being unable to pay our bills.

Choose an appropriate response.

INTERVIEWER: From what you say, you have two major problems. Perhaps the one we should look at first is your inability to pay your bills. (Go to 14.20)

INTERVIEWER: You have two urgent problems: finding a job and paying your bills. Paying your bills seems to be your greater concern. To what extent should we deal with that concern first? (Go to 14.21)

INTERVIEWER: It's often better to consider the easiest problem first. Maybe we should look for ways to get you and your wife out socially. (Go to 14.22)

14.15

CLIENT: I have a lot of difficulties because I'm out of work. If I had a job, a lot of my problems would be settled. I'd really like to get back into real estate, but no one wants to hire me.

Choose an appropriate response.

INTERVIEWER: The real estate market is slow right now. Do you have to go into that business? (Go to 14.23)

INTERVIEWER: You seem really determined to go into real estate. Perhaps I'm wrong, but I'm getting the feeling that it's difficult for you to consider any other alternatives. (Go to 14.24)

INTERVIEWER: It sounds as though you lost your license for some reason. Maybe that's why you're having trouble getting back into real estate. (Go to 14.25)

14.16

CLIENT: That's why I contacted you. I can't go on like this. I've got debts building up, and they have to be paid. It's beginning to create problems at home. I'm sure that, if I got a job, my problems would be settled. The job's the key.

Choose an appropriate response.

INTERVIEWER: From what you say, it seems that you're overreacting. There's no need to get so upset. (Go to 14.26)

INTERVIEWER: Since you have so many debts, maybe you should see a debt counselor. (Go to 14.27)

INTERVIEWER: Getting a job is vital if you're going to hold things together. What else do you think you could do to find a job? (Go to 14.28)

14.17 Your answer: *So, you have two urgent problems. Could you help me to understand which one is more pressing?*

Correct. This restatement of content, followed by an open question, constitutes a structuring response that will help the client deal with his problems in order of their priority. Go to 14.33.

14.18 Your answer: *Everyone seems to be in the same situation these days. Is there anyone you could borrow money from?*

The client has sought your assistance in a crisis and is in need of your understanding and support. He has probably exhausted all possible short-term solutions, including loans. Your response could negate the effort he made in contacting you. Return to 14.13 and try again.

14.19 Your answer: *I get the feeling that you really haven't been trying hard enough to find a job. What things did you say you'd done so far?*

You've berated the client for not doing enough to find a job, and then you've suggested that you can't remember what he has done in this regard. Also, you've neglected to discuss the order in which his problems should be considered. Return to 14.13 and try again.

14.20 Your answer: *From what you say, you have two major problems. Perhaps the one we should look at first is your inability to pay your bills.*

Correct. Your restatement indicates that you're following the client. Furthermore, you've checked with the client on the appropriateness of attending to his financial problem first. Go to 14.32.

14.21 Your answer: *You have two urgent problems: finding a job and paying your bills. Paying your bills seems to be your greater concern. To what extent should we deal with that concern first?*

Correct. Your paraphrase demonstrates to the client that you've heard his comments, and your open question structures the situation for him to consider the order in which his problems should be examined. Go to 14.31.

14.22 Your answer: *It's often better to consider the easiest problem first. Maybe we should look for ways to get you and your wife out socially.*

The client has said that his problems would be solved if he could find a job. His lack of a social life has been alluded to, but only in terms of a side effect of his major problem. In times of crisis, major problems are the focus of attention. At other times, the interviewer and the client decide on the order in which problems are to be considered. Return to 14.14 and try again.

14.23 Your answer: *The real estate market is slow right now. Do you have to go into that business?*

This closed question makes you responsible for the direction of the interview.

Your response should keep the discussion open and give the client this responsibility. Return to 14.15 and try again.

14.24 **Your answer:** *You seem really determined to go into real estate. Perhaps I'm wrong, but I'm getting the feeling that it's difficult for you to consider any other alternatives.*
Correct. You have confronted the client with the discrepancy between his need to find a job and his reluctance to consider a wider choice of alternatives. Go to 14.30.

14.25 **Your answer:** *It sounds as though you lost your license for some reason. Maybe that's why you're having trouble getting back into real estate.*
The client has given you no information about losing his license. In making an interpretation, you must ensure that you have correctly identified the client's implicit message and then present the interpretation in a tentative manner. Return to 14.15 and try again.

14.26 **Your answer:** *From what you say, it seems that you're overreacting. There's no need to get so upset.*
Your response denies the client the right to express and explore his feelings and may cause him to think that you're unable to understand his problem. Return to 14.16 and try again.

14.27 **Your answer:** *Since you have so many debts, maybe you should see a debt counselor.*
Providing information is an important skill in interviewing, but the information must be provided at an appropriate time. The client's major concern is a job. Return to 14.16 and try again.

14.28 **Your answer:** *Getting a job is vital if you're going to hold things together. What else do you think you could do to find a job?*
Correct. You've accurately reflected the client's feelings of urgency about finding a job and interpreted this as a reasonable goal. Your open question is a structuring response that will initiate the preparation for action stage of the interview. Go to 14.29.

14.29 CLIENT: Well, as I mentioned, there's real estate, but there really aren't any openings there. I hated working in a factory, and I can't see doing it again. My wife has suggested that I look for something new, even if it means retraining. I've thought a lot about that, and it seems to be the best route to take.

Choose an appropriate response.

INTERVIEWER: As you look at the options that are available to you, retraining seems to be the most workable. (Go to 14.34)

INTERVIEWER: Do you think it's wise to retrain just because your wife suggested it? (Go to 14.35)

INTERVIEWER: I find it hard to believe that there aren't any openings in real estate. (Go to 14.36)

14.30

CLIENT: I really would like to get into real estate, but I'm not having much success. I've had to look at other alternatives. I could go back into factory work, but I hated that. I'd be willing to take a job in which I'd move around and meet people, but I don't think I'm qualified for anything other than real estate. The whole situation is sort of frightening.

Choose an appropriate response.

INTERVIEWER: Have you tried applying for outdoor jobs? You might be able to find work as a gardener. There are a number of nurseries and lawn-maintenance firms in this area. (Go to 14.37)

INTERVIEWER: There's a certain amount of fear involved in thinking about what to do. The jobs you can do seem to be impossible to find, and any other job represents an unknown. (Go to 14.38)

INTERVIEWER: If you keep your fear under control, you'll be more likely to find a job. (Go to 14.39)

14.31

CLIENT: Well, I guess we have to look at that first. I hate to think of what will happen if I don't pay our bills this week. I've thought of a number of alternatives, but the only one that seems reasonable is to go on unemployment. But I'm not sure you're here to help me with that.

Choose an appropriate response.

INTERVIEWER: I'm a little concerned that you've ruled out a loan so hastily. It's your only solution right now. (Go to 14.40)

INTERVIEWER: You seem to have some reservations about going on unemployment. Could you tell me your thoughts about that? (Go to 14.41)

INTERVIEWER: You're asking me whether I can help you to investigate the possibility of applying for unemployment benefits. (Go to 14.42)

14.32

CLIENT: Yes. I really have to do something about that. By the end of next week, I need to find enough money to pay the rent and utilities, buy groceries, and buy shoes for the children. And then there's next month.

Choose an appropriate response.

INTERVIEWER: You haven't told me anything about your family. Perhaps you could tell me a little bit about them now. (Go to 14.43)

INTERVIEWER: I don't think there's any way to solve your problems in 1 week. (Go to 14.44)

INTERVIEWER: You seem to have a reasonable goal, but one that must be creating a lot of pressure for you. What do you think you could do to achieve that goal? (Go to 14.45)

14.33

CLIENT: They're both important, but I came to talk to you because I don't even have enough money to pay the rent next week.

Choose an appropriate response.

INTERVIEWER: Your most urgent problem is a lack of money. Could you tell me whether you're buying your house? (Go to 14.46)

INTERVIEWER: We should consider your money problem first. What do you need to do in order to resolve your financial crisis? (Go to 14.47)

INTERVIEWER: We should attend to your money problems first. Perhaps you should find accommodations with your relatives. (Go to 14.48)

14.34 Your answer: *As you look at the options that are available to you, retraining seems to be the most workable.*
Correct. This restatement enables you and the client to verify or modify the plan that is taking shape. Go to 14.54.

14.35 Your answer: *Do you think it's wise to retrain just because your wife suggested it?*
The client told you that, after his wife suggested retraining, he gave the matter careful thought. By ignoring this part of the client's message, you've failed to use the most basic interviewing skill—attending. Furthermore, your closed question makes you responsible for the reestablishment of the discussion. Return to 14.29 and try again.

14.36 Your answer: *I find it hard to believe that there aren't any openings in real estate.*
If you feel that it is important to discuss this topic, you should use an open question or a reflection of content to initiate such a discussion. However, the client has placed considerable stress on the possibility of retraining; you should pursue this topic. Return to 14.29 and try again.

14.37 Your answer: *Have you tried applying for outdoor jobs? You might be able to find work as a gardener. There are a number of nurseries and lawn-maintenance firms in this area.*
This closed question, followed by advice, is inappropriate at this point in the interview. Help the client explore a variety of action plans and select the most appropriate one. Your response should stimulate his exploration of the alternatives available to him. Return to 14.30 and try again.

14.38 Your answer: *There's a certain amount of fear involved in thinking about what to do. The jobs you can do seem to be impossible to find, and any other job represents an unknown.*
Correct. This reflection of feeling and content encourages the client to discuss his apprehensions, bring them under control, and continue his exploration of the alternatives available to him. Go to 14.53.

14.39 Your answer: *If you keep your fear under control, you'll be more likely to find a job.*
This type of well-meaning advice is generally ineffective. The client has told you that he is afraid. He will need your understanding and support to de-

velop the ability to cope with his fears and find suitable work. Return to 14.30 and try again.

14.40	**Your answer:**	*I'm a little concerned that you've ruled out a loan so hastily. It's your only solution right now.*

Your task is to help the client explore his problem and select an action plan that is acceptable to him. Any attempt on your part to prescribe a solution is inappropriate. Return to 14.31 and try again.

14.41	**Your answer:**	*You seem to have some reservations about going on unemployment. Could you tell me your thoughts about that?*

Correct. You've reflected the client's concern about applying for unemployment benefits and invited him to explore this concern with you. Go to 14.51.

14.42	**Your answer:**	*You're asking me whether I can help you to investigate the possibility of applying for unemployment benefits.*

Correct. This communication of immediacy and interpretation identifies the client's concern about his relationship with you. This response will help him become aware of his concern and work toward its resolution. Go to 14.52.

14.43	**Your answer:**	*You haven't told me anything about your family. Perhaps you could tell me a little bit about them now.*

The client is trying to solve a major financial difficulty; your response distracts him from this task. Your failure to follow the client suggests that you aren't interested in the problem he is discussing. The client's progress toward solving his problem is likely to be impeded by your response. Return to 14.32 and try again.

14.44	**Your answer:**	*I don't think there's any way to solve your problems in 1 week.*

Your concerned opinion is inappropriate here and will do little to help the client solve his problem. Try to generate alternative action plans with the client by discussing possible solutions with him. Return to 14.32 and try again.

14.45	**Your answer:**	*You seem to have a reasonable goal, but one that must be creating a lot of pressure for you. What do you think you could do to achieve that goal?*

Correct. You've confirmed the legitimacy of the client's goal in a supportive manner and structured the situation so he can consider possible alternatives to his problem. Go to 14.50.

14.46	**Your answer:**	*Your most urgent problem is a lack of money. Could you tell me whether you're buying your house?*

Although you've agreed with your client concerning the order in which to deal with his problems, your closed question changes the topic. At this point, you need to use a more open response that will encourage the client to engage in the goal-formulation process. Return to 14.33 and try again.

14.47 Your answer: *We should consider your money problems first. What do you need to do in order to resolve your financial crisis?*

Correct. This structuring response demonstrates that you agree with the client. Your open question encourages him to translate his problem into a goal. Go to 14.49.

14.48 Your answer: *We should attend to your money problems first. Perhaps you should find accommodations with your relatives.*

Although you've agreed with the client concerning the high priority of his financial problems, you've attempted to solve his problem on your own and prematurely. You've neglected goal formulation—a prerequisite to problem solving. Return to 14.33 and try again.

14.49

CLIENT: I need enough money to pay the rent and utilities, buy groceries, and buy shoes for the children next week. And it will be the same sort of thing every month.

INTERVIEWER: That, seems reasonable. . . . Those things are necessities. What do you think you can do to pay for them?

CLIENT: Well, it would be nice if I had a job, but I doubt that I'll find one right away. I could go on unemployment, but that's like going on welfare. I could get a loan, but I have no idea when I could pay it back.

Choose an appropriate response.

INTERVIEWER: There's nothing wrong with going on unemployment. You just have to go down to the office and apply for it. (Go to 14.55)

INTERVIEWER: Even though your financial situation is desperate, you're reluctant to apply for unemployment benefits. It seems that applying for them runs contrary to your values. (Go to 14.56)

INTERVIEWER: You say that you could get a loan. Do you have a good credit rating? (Go to 14.57)

14.50

CLIENT: Well, it would be nice if I had a job, but I doubt that I'll find one right away. I could go on unemployment, but that's like going on welfare. I could get a loan, but I have no idea when I could pay it back.

Choose an appropriate response.

INTERVIEWER: Even though your financial situation is desperate, you're reluctant to apply for unemployment benefits. It seems that applying for them runs contrary to your values. (Go to 14.58)

INTERVIEWER: Maybe you should get a small loan for a month and then try really hard during that month to find a job. (Go to 14.59)

INTERVIEWER: Have you ever had a loan? (Go to 14.60)

14.51

CLIENT: I don't know that it's a problem really. I just can't go on unemployment benefits. It's like going on welfare.

Choose an appropriate response.

INTERVIEWER: Since you have such strong feelings about this, we'll have to search for another alternative. (Go to 14.61)

INTERVIEWER: In your situation, you should have been on unemployment months ago. (Go to 14.62)

INTERVIEWER: Even though your financial situation is desperate, you're reluctant to apply for unemployment benefits. It seems that applying for them runs contrary to your values. (Go to 14.63)

14.52

CLIENT: Well, that's not really my concern. I know you can help me. I just don't want to talk about going on unemployment. It's just like going on welfare.

Choose an appropriate response.

INTERVIEWER: Since you have such confidence in me, I'm sure you'll follow my advice and apply for unemployment benefits as soon as possible. (Go to 14.64)

INTERVIEWER: Even though your financial situation is desperate, you're reluctant to apply for unemployment benefits. It seems that applying for them runs contrary to your values. (Go to 14.65)

INTERVIEWER: I have to stress that you have a right to unemployment benefits and that it's not like going on welfare. (Go to 14.66)

14.53

CLIENT: But I don't want just any job. I've had this uncertainty. What I want is a secure job, and that means I'll have to retrain. I just don't know what area to enter or what I could do.

INTERVIEWER: (summarizing and providing information) What I hear you saying is that you feel you've reached a turning point in your work situation, and that, before you make any decisions, you need to examine your skills and your possibilities very seriously. It's not part of my job to become involved in career planning, but I can refer you to a vocational counselor. The appointment would be for tomorrow.

CLIENT: I've avoided doing that for a long time, but I know I should. I have to begin somewhere. I'll be there whenever you say.

Choose an appropriate response.

INTERVIEWER: To summarize, then, you've decided to explore your career possibilities with our vocational counselor, and you'll see him in the morning. I wonder if there's anything else you'd like to discuss before we conclude. (Go to 14.67)

INTERVIEWER: To summarize, then, you've decided to explore your career possibilities with our vocational counselor, and you'll see him in the morning. I guess we've covered everything. I'm sure that your problems will be sorted out now. (Go to 14.68)

INTERVIEWER: To summarize, then, you've decided to explore your career possibilities with our vocational counselor, and you'll see him in the morning. Perhaps we should discuss the kinds of jobs you might be interested in. (Go to 14.69)

14.54   CLIENT: I'm beginning to feel that way more and more. My biggest problem is that I don't know how to go about it.

INTERVIEWER: (paraphrasing and providing information) So, you need some help in arranging a retraining program. It seems essential that you see a vocational counselor as soon as possible. If you agree, I can arrange an appointment for tomorrow morning.

CLIENT: That sounds fine to me.

Choose an appropriate response.

INTERVIEWER: Well, I think that we've dealt with all your problems. Make sure that you show up for your appointment in the morning. (Go to 14.70)

INTERVIEWER: To summarize, then, you've decided to explore your career possibilities with our vocational counselor, and you'll see him in the morning. I wonder if there's anything else you'd like to discuss before we conclude. (Go to 14.71)

INTERVIEWER: You've been out of work for some time, and you can't find a job in your former occupation. Your wife has told you that you should try something new. You want to hold your marriage together, so you're going along with this. I think everything is under control now. We can conclude at this point. (Go to 14.72)

14.55 Your answer: *There's nothing wrong with going on unemployment. You just have to go down to the office and apply for it.*
At this point, the client needs help confronting his feelings about using the one financial resource available to him, not advice and instructions. Return to 14.49 and try again.

14.56 Your answer: *Even though your financial situation is desperate, you're reluctant to apply for unemployment benefits. It seems that applying for them runs contrary to your values.*
Correct. By confronting the client with his financial need and the resources that are available to him, you help him examine his feelings about applying for unemployment benefits. Go to 14.74.

14.57 Your answer: *You say that you could get a loan. Do you have a good credit rating?*
Your closed question prohibits discussion of this option. In order to explore this possibility, respond in a way that leads to an open discussion. However, the discrepancy between the client's financial situation and his reluctance to apply for unemployment benefits requires confrontation at this point. Return to 14.49 and try again.

14.58 Your answer: *Even though your financial situation is desperate, you're reluctant to apply for unemployment benefits. It seems that applying for them runs contrary to your values.*
Correct. By confronting the client with his financial need and the resources that are available to him, you help him examine his feelings about applying for unemployment benefits. Go to 14.74.

14.59	**Your answer:**	*Maybe you should get a small loan for a month and then try really hard during that month to find a job.*

The client has generated a number of alternative courses of action. You need to confront him with any discrepancies in his statements and then discuss the relative merits of each course of action. Return to 14.50 and try again.

14.60	**Your answer:**	*Have you ever had a loan?*

This closed question calls for a simple "yes" or "no" answer and places the responsibility for the discussion on you. The client is working toward a solution of his financial problems; your response should encourage him to continue his problem solving. Return to 14.50 and try again.

14.61	**Your answer:**	*Since you have such strong feelings about this, we'll have to search for another alternative.*

Because the client's strong feelings may be based on a misunderstanding or a false assumption, you should confront him with his feelings so he can explore and resolve them. Return to 14.51 and try again.

14.62	**Your answer:**	*In your situation, you should have been on unemployment months ago.*

This confrontation is much too direct. It will either put the client on the defensive or cause him to avoid the topic altogether. Confrontations should be tentative, not aggressive. When you confront tentatively, the client is more likely to explore feelings and behavior and arrive at a resolution. Return to 14.51 and try again.

14.63	**Your answer:**	*Even though your financial situation is desperate, you're reluctant to apply for unemployment benefits. It seems that applying for them runs contrary to your values.*

Correct. By confronting the client with his financial need and the resources that are available to him, you help him examine his feelings about applying for unemployment benefits. Go to 14.74.

14.64	**Your answer:**	*Since you have such confidence in me, I'm sure you'll follow my advice and apply for unemployment benefits as soon as possible.*

It's inappropriate for you to use your authority and your position to influence the client. Instead, you should facilitate discussion of his feelings about unemployment benefits. Return to 14.52 and try again.

14.65	**Your answer:**	*Even though your financial situation is desperate, you're reluctant to apply for unemployment benefits. It seems that applying for them runs contrary to your values.*

Correct. By confronting the client with his financial need and the resources that are available to him, you help him examine his feelings about applying for unemployment benefits. Go to 14.74.

14.66	**Your answer:**	*I have to stress that you have a right to unemployment benefits and that it's not like going on welfare.*

By emphasizing your own views, you've denied the client the right to his own feelings. He is unlikely to change his views until you help him explore his feelings and the assumptions that underlie them. Return to 14.52 and try again.

14.67 Your answer: *To summarize, then, you've decided to explore your career possibilities with our vocational counselor, and you'll see him in the morning. I wonder if there's anything else you'd like to discuss before we conclude.*

Correct. Your restatement of the decision that you and the client have made provides him with structure for his upcoming appointment. Also, you've given him an opportunity to discuss other problems. Go to 14.73.

14.68 Your answer: *To summarize, then, you've decided to explore your career possibilities with our vocational counselor, and you'll see him in the morning. I guess we've covered everything. I'm sure that your problems will be sorted out now.*

Your restatement is appropriate, but you've attempted to terminate the interview prematurely. In his initial statement, the client said that he has financial problems; your final statement assumes that these problems were resolved when the referral to the vocational counselor was made. This may be the case, but you need to check this out with the client before terminating the interview. Return to 14.53 and try again.

14.69 Your answer: *To summarize, then, you've decided to explore your career possibilities with our vocational counselor, and you'll see him in the morning. Perhaps we should discuss the kinds of jobs you might be interested in.*

Since you've referred the client to a vocational counselor, your restatement is appropriate, but your attempt to discuss his job situation is not. You might invite him to explore other problem areas with you. If he declines this invitation, you should terminate the interview. Return to 14.53 and try again.

14.70 Your answer: *Well, I think that we've dealt with all your problems. Make sure that you show up for your appointment in the morning.*

This response is premature and abrupt. Before you terminate the interview, you should give the client a chance to discuss any other problems he may have. If the client doesn't have any other problems to discuss, you should terminate the interview. Return to 14.54 and try again.

14.71 Your answer: *To summarize, then, you've decided to explore your career possibilities with our vocational counselor, and you'll see him in the morning. I wonder if there's anything else you'd like to discuss before we conclude.*

Correct. Your restatement of the decision that you and the client have made provides him with structure for his upcoming appointment. Also, you've given him an opportunity to discuss other problems. Go to 14.73.

14.72 Your answer: *You've been out of work for some time, and you can't find a job in your former occupation. Your wife has told you that you should try something new. You want to hold your marriage together, so you're going along with this. I think everything is under control now. We can conclude at this point.*

The client has told you that the decision to retrain was his. Your inaccurate summarization deflects him from focusing on the career plan he is attempting to formulate. Moreover, both you and the client must decide when to conclude the interview. Return to 14.54 and try again.

14.73

CLIENT: Well . . . there is one problem. How are we going to pay our bills next week? There just isn't any money left.

Choose an appropriate response.

INTERVIEWER: Your financial situation is critical, and you don't know what to do about it. (Go to 14.75)

INTERVIEWER: Couldn't you secure a loan to cover you until you find a job? (Go to 14.76)

INTERVIEWER: That must be making things very difficult for you. (Go to 14.77)

14.74

CLIENT: When you put it that way, I guess I can see why I should apply. It's just that, in my family, that sort of thing was looked down on. But then, nobody in my family was ever in this position. My father always had a good job, so how could he know? I guess I'd better apply. My wife will be pleased. But if I do that, it's even more important that I find a job as soon as possible.

Choose an appropriate response.

INTERVIEWER: I get the feeling that there's some friction between you and your wife. Is that right? (Go to 14.78)

INTERVIEWER: Your decision to apply for unemployment solves your immediate financial crisis but makes your desire to find work even stronger. (Go to 14.79)

INTERVIEWER: I'm glad to hear you say that you'll apply. That's one problem under control. Let's move on now to discuss the job issue. How are you going to go about finding a job? (Go to 14.80)

14.75 Your answer: *Your financial situation is critical, and you don't know what to do about it.*
Correct. By restating the content of the client's message, you verify that you've understood him and facilitate further discussion of his problem. Go to 14.83.

14.76 Your answer: *Couldn't you secure a loan to cover you until you find a job?*
This closed question is inappropriate at this point in the interview. As the client begins to talk about a problem, you need to make a response that will facilitate discussion. An open question or a reflective response is an appropriate stimulus for such a discussion. Return to 14.73 and try again.

14.77 Your answer: *That must be making things very difficult for you.*
Correct. By reflecting feeling, you encourage the client to tell you about his difficulty and provide him with support. Go to 14.84.

14.78 Your answer: *I get the feeling that there's some friction between you and your wife. Is that right?*
There may be friction between the client and his wife; however, he hasn't raised this as a major problem. This interpretation, even if accurate, distracts him from his present line of concern and demonstrates poor attending. Also, your closed question will require you to direct the discussion once he has answered "yes" or "no." Return to 14.74 and try again.

14.79 Your answer: *Your decision to apply for unemployment solves your immediate financial crisis but makes your desire to find work even stronger.*
Correct. This accurate restatement of the client's feelings about his situation shows him that you are interested in what he is saying. Go to 14.81.

14.80 Your answer: *I'm glad to hear you say that you'll apply. That's one problem under control. Let's move on now to discuss the job issue. How are you going to go about finding a job?*
Correct. Your expression of feeling is appropriate, and you've structured the interview in such a way that the client's other concern—finding a job—becomes the focus of discussion. Go to 14.82.

14.81

CLIENT: I just have to find work now. When I called you, I thought you might be able to help me make some plans. One of the things I had thought about was retraining to work in another field, but I don't know how to go about it or how I would live if I did that. Can you help me?

Choose an appropriate response.

INTERVIEWER: I wonder if retraining is really appropriate, given your age and responsibilities. It might be better to look for a job for which you are already trained. (Go to 14.85)

INTERVIEWER: Do you think you should enter retraining? What does your wife think about it? (Go to 14.86)

INTERVIEWER: You really have given this some thought. Maybe we should set up an appointment for you to come in tomorrow and speak with a vocational counselor. How does that idea sound to you? (Go to 14.87)

14.82

CLIENT: Well, it's really a problem. I thought that when I contacted you, you'd give me some ideas. I've exhausted the list of jobs I'm trained for. There's just nothing in real estate. There's no heavy industry here, not that I want to get into that again. Do you think I would be able to find work in another field altogether if I retrained? I've tried to look at what I might do.

Choose an appropriate response.

INTERVIEWER: You say you've exhausted the list of jobs you're trained for. I don't really think that's possible. There must be jobs you don't even know about. I know it's difficult, but you have to keep trying. (Go to 14.88)

INTERVIEWER: I believe your chances of finding work in another area are very poor. Thousands of people are looking for work, and as a result, employers have many applicants to choose from. You can be sure that a person who has experience will be preferred. (Go to 14.89)

INTERVIEWER: It seems as though you've been giving this serious consideration. I think you should see a vocational counselor. How do you feel about that? (Go to 14.90)

14.83

CLIENT: It is critical. As far as I can see, there's only one thing to do. I'll go on unemployment. But up until now, I've refused to do that. I suppose that as long as I'm doing something about retraining for a job, it will be all right.

Choose an appropriate response.

INTERVIEWER: Now that you've decided to take some steps to find a job, you feel a little more comfortable about applying for unemployment benefits. It's important to you to earn the money you receive. (Go to 14.91)

INTERVIEWER: I don't understand why you feel that way. We all pay unemployment insurance, and we're all entitled to it, including you. I don't know why you aren't on unemployment already. (Go to 14.92)

INTERVIEWER: I think you've been unduly harsh toward your family by refusing to go on unemployment. They have to be considered, too. (Go to 14.93)

14.84

CLIENT: Yes. It's making things very difficult. I guess some of it is due to my stubbornness about applying for unemployment benefits. I've always looked down on people who are on unemployment. I thought that they were lazy. But now that it's me, it's hard to admit that I've been wrong all this time. . . . I do need the money.

Choose an appropriate response.

INTERVIEWER: I once thought that people on unemployment were a little lazy. (Go to 14.94)

INTERVIEWER: Where did you get that idea? Everyone is entitled to unemployment benefits. (Go to 14.95)

INTERVIEWER: Until now, you've associated receiving unemployment benefits with being lazy. It's hard to change your views, even when you're involved. (Go to 14.96)

14.85 Your answer: *I wonder if retraining is really appropriate, given your age and responsibilities. It might be better to look for a job for which you are already trained.*
If this decision is made, it should follow from a careful exploration of the client's abilities and work potential. It shouldn't result from an opinion based on limited information. Return to 14.81 and try again.

14.86 Your answer: *Do you think you should enter retraining? What does your wife think about it?*
You've asked the client one question after another without giving him an opportunity to respond. Moreover, you've changed topics. If this information is important to the client he will bring it up when he feels that it is relevant. Return to 14.81 and try again.

14.87 Your answer: *You really have given this some thought. Maybe we should set up an appointment for you to come in tomorrow and speak with a vocational counselor. How does that idea sound to you?*

Correct. In your summary sentence, you've indicated that you realize how much effort the client has expended. You then structured the situation so that the client was provided with information and an appropriate action plan. Finally, you've invited him to discuss his reaction. Go to 14.100.

14.88 Your answer: *You say you've exhausted the list of jobs you're trained for. I don't really think that's possible. There must be jobs you don't even know about. I know it's difficult, but you have to keep trying.*

Since the client feels that he has exhausted his possibilities, your statement will give him the impression that you don't understand his situation. This could cause him to withdraw from his relationship with you. To help him, you need to explore his problem with him so that you understand his situation. Return to 14.82 and try again.

14.89 Your answer: *I believe your chances of finding work in another area are very poor. Thousands of people are looking for work, and as a result, employers have many applicants to choose from. You can be sure that a person who has experience will be preferred.*

This pessimistic response is not appropriate at this point in the interview. If you feel it's important to discuss the consequences of the client's action plan, you should ask him to explore those consequences. An open question would be appropriate in that case. However, because of the complexity of the job situation, this discussion is best left to another time. Return to 14.82 and try again.

14.90 Your answer: *It seems as though you've been giving this serious consideration. I think you should see a vocational counselor. How do you feel about that?*

Correct. Your brief restatement of what the client has said, followed by information and a proposed action plan, will help him deal with his problem. Also, you've invited him to respond to your suggestion. Go to 14.99.

14.91 Your answer: *Now that you've decided to take some steps to find a job, you feel a little more comfortable about applying for unemployment benefits. It's important to you to earn the money you receive.*

Correct. This reflection of feeling and content identifies the important aspects of the client's problem. You've encouraged him to explore his problem further. Go to 14.98.

14.92 Your answer: *I don't understand why you feel that way. We all pay unemployment insurance, and we're all entitled to it, including you. I don't know why you aren't on unemployment already.*

It's inappropriate for you to impose your biases on the client. You should stimulate discussion of this issue so that he can reach his own conclusions. Return to 14.83 and try again.

14.93 Your answer: *I think you've been unduly harsh toward your family by refusing to go on unemployment. They have to be considered, too.*
This judgmental statement could destroy the rapport between you and the client. It will not help him explore his feelings or resolve his problems. Return to 14.83 and try again.

14.94 Your answer: *I once thought that people on unemployment were a little lazy.*
This inappropriate self-disclosure might reinforce the client's bias. Self-disclosure should facilitate rather than inhibit the client's action. Return to 14.84 and try again.

14.95 Your answer: *Where did you get that idea? Everyone is entitled to unemployment benefits.*
Considering the client's views about unemployment benefits, it is inappropriate to impose your opinion on him. Your response should demonstrate that you are aware of the difficulties the client is experiencing and encourage him to continue discussion of this topic. Return to 14.84 and try again.

14.96 Your answer: *Until now, you've associated receiving unemployment benefits with being lazy. It's hard to change your views, even when you're involved.*
Correct. Your accurate reflection of the client's concern identifies the source of his difficulties and encourages him to try to resolve them. Go to 14.97.

14.97

CLIENT: It's hard to change your views. I don't like being wrong. I guess I was wrong, and it's time to admit it. Then I can go down and apply for my benefits. When I do that, I'll solve my immediate problems. And I'll see the vocational counselor tomorrow.

Choose an appropriate response.

INTERVIEWER: It sounds as though you maintain very mixed views. Could you tell me about any other instances in which your beliefs created problems for you? (Go to 14.101)

INTERVIEWER: You're comfortable with the decisions you've made, and you feel you can go on from here. Now that you have a plan, I think this is a good time to close. (Go to 14.102)

INTERVIEWER: It's good to see you taking some action. Before we close, I want to stress the importance of following through with those plans. We won't be able to do this sort of thing again, so do follow through. (Go to 14.103)

14.98

CLIENT: Yes. I'm more comfortable when I earn my money. But I guess there are other things that are more important, like taking care of my family and paying the bills. I'll apply for unemployment right away, and then it will be done.

Choose an appropriate response.

INTERVIEWER: Before you do that, maybe we should talk about your family some more. (Go to 14.104)

INTERVIEWER: That's a good idea. As soon as you've applied, you'll feel great. I'm sure you'll realize then that you've made a great deal of fuss about nothing. (Go to 14.105)

INTERVIEWER: It will be difficult, but it will be over. Now, just to review your plans, you're coming to see the vocational counselor in the morning, and you're going to apply for unemployment benefits right away. Perhaps this would be an appropriate time for us to conclude the interview. If anything comes up, feel free to get back to me. (Go to 14.106)

14.99

CLIENT: I think I'd prefer to talk it through with you, but if you feel I should see the vocational counselor, I will.

Choose an appropriate response.

INTERVIEWER: I get the feeling that you find it difficult to meet people. You really shouldn't let your problems get you down like that. (Go to 14.107)

INTERVIEWER: It's difficult to start all over again. I'd be hesitant, just as you are. But a vocational counselor will be able to help you more than I can. I'll set up an appointment for you, and if you have any concerns in the meantime, please feel free to get back to me. (Go to 14.108)

INTERVIEWER: Please don't misunderstand me. I would like to discuss your problem with you, but you should see the vocational counselor. I have to meet with another client now. (Go to 14.109)

14.100

CLIENT: Well . . . if I'm going to sort this out, I suppose I need to do it properly this time. I don't ever want to be mixed up like this again. You set it up, and I'll be there.

Choose an appropriate response.

INTERVIEWER: Let's just briefly review your plans. You're going to apply for unemployment benefits as soon as possible, and you're coming in tomorrow to discuss your job situation with the vocational counselor. If you have any further difficulties, feel free to get back to me. (Go to 14.110)

INTERVIEWER: Mixed up. What do you mean by "mixed up"? I thought that your problem was settled now. Is there something else? (Go to 14.111)

INTERVIEWER: You must be relieved to have made these decisions. How do you think you should behave in the future if a situation like this develops? (Go to 14.112)

14.101 Your answer: *It sounds as though you maintain very fixed views. Could you tell me about any other instances in which your beliefs created problems for you?*

The client has told you that he has the situation under control and is comfortable with the action plans the two of you have formulated. It's inappropriate to initiate a discussion of his past experiences at this time. You should focus on the relevant issue the client has brought to your attention—the termination of the interview. Return to 14.97 and try again.

14.102 Your answer:

You're comfortable with the decisions you've made, and you feel you can go on from here. Now that you have a plan, I think this is a good time to close.

Correct. Your restatement accurately reflects the client's message, and your structuring response shows that you agree that it is appropriate to close the interview.

In reaching this point, you've used a variety of interviewing skills. Now return to 14.1 and choose another appropriate response. You've already chosen 14.4, so don't choose that response. If you've already completed your second sequence of responses, go to 14.113.

14.103 Your answer:

It's good to see you taking some action. Before we close, I want to stress the importance of following through with those plans. We won't be able to do this sort of thing again, so do follow through.

The client hasn't indicated that he won't follow through with the decisions he has made. At this point, your response should merely close the interview. Return to 14.97 and try again.

14.104 Your answer:

Before you do that, maybe we should talk about your family some more.

This response gives the client no structure within which to reply; therefore, it will probably confuse him. In responding to a client, attend to his message and provide a coherent structure. Return to 14.98 and try again.

14.105 Your answer:

That's a good idea. As soon as you've applied, you'll feel great. I'm sure you'll realize then that you've made a great deal of fuss about nothing.

It's more likely that the client will continue to feel discomfort until he has found suitable work. You need to identify and reflect feelings accurately. Return to 14.98 and try again.

14.106 Your answer:

It will be difficult, but it will be over. Now, just to review your plans. You're coming to see the vocational counselor in the morning, and you're going to apply for unemployment benefits right away. Perhaps this would be an appropriate time for us to conclude the interview. If anything comes up, feel free to get back to me.

Correct. Following a brief reflective statement, you've summarized the decisions that you and the client have made and provided a support system for him.

In reaching this point, you've used a variety of interviewing skills. Now return to 14.1 and choose another appropriate response. You've already chosen 14.4, so don't choose that response. If you've already completed your second sequence of responses, go to 14.113.

14.107 Your answer: *I get the feeling that you find it difficult to meet people. You really shouldn't let your problems get you down like that.*

You should follow the client carefully and focus only on relevant information, not on assumptions or interpretations. The client hasn't indicated that his difficulty in changing interviewers is associated with problems in meeting people. It's more likely that he finds it difficult to discuss his problem and is reacting to the necessity of explaining his situation again. Return to 14.99 and try again.

14.108 Your answer: *It's difficult to start all over again. I'd be hesitant, just as you are. But a vocational counselor will be able to help you more than I can. I'll set up an appointment for you, and if you have any concerns in the meantime, please feel free to get back to me.*

Correct. By accurately reflecting feeling and by self-disclosing, you help the client cope with the difficulty of changing interviewers. Your structuring response provides an explanation for this change. You've also provided a support system for the client.

In reaching this point, you've used a variety of interviewing skills. Now return to 14.1 and choose another appropriate response. You've already chosen 14.3, so don't choose that response. If you've already completed your second sequence of responses, go to 14.113.

14.109 Your answer: *Please don't misunderstand me. I would like to discuss your problem with you, but you should see the vocational counselor. I have to meet with another client now.*

This abrupt response suggests that the client has been an imposition. Leave the relationship open so the client will feel free to contact you if he requires your help in the future. This response is more likely to make him think twice about contacting you again. Return to 14.99 and try again.

14.110 Your answer: *Let's just briefly review your plans. You're going to apply for unemployment benefits as soon as possible, and you're coming in tomorrow to discuss your job situation with the vocational counselor. If you have any further difficulties, feel free to get back to me.*

Correct. This structuring response briefly reviews what has occurred during the interview and offers the client the option of returning to see you.

In reaching this point, you've used a variety of interviewing skills. Now return to 14.1 and choose another appropriate response. You've already chosen 14.3, so don't choose that response. If you've already completed your second sequence of responses, go to 14.113.

14.111 Your answer: *Mixed up. What do you mean by "mixed up"? I thought your problem was settled now. Is there something else?*

It's inappropriate to make assumptions based on insufficient information. Instead, help the client explore a problem during the interview so that more information is available. However, because you are moving toward the termination of the interview, this strategy is not appropriate at this time. If the client is concerned about another issue, arrange to meet with him again. Return to 14.100 and try again.

14.112 Your answer:	*You must be relieved to have made these decisions. How do you think you should behave in the future if a situation like this develops?*

This open question will initiate discussion when the interview is moving toward its conclusion. At this point, your response should summarize the interview and bring it to a close. Return to 14.100 and try again.

14.113

The material in this chapter is intended to help you integrate the skills covered in this book. As you have observed, a variety of skills can be used in any one situation. The manner in which you adapt and develop these skills will reflect your own personal interviewing style. To acquire a relaxed, natural, and facilitative style, you must practice these skills, and to maintain your competence, you must review your interviews periodically. One way of doing this is to tape an interview and then identify the types of responses you use and their effects. As you gain experience, you will develop skills to complement those you've learned here, establishing your own effective style.

Points to Remember about Integrating the Skills

1. WHEN USING SKILLS IN AN INTERVIEW:
   a. Use each skill expertly.
   b. Structure the interview appropriately.
   c. Respond in a way that is both facilitative and comfortable.
   d. Combine the skills as required.
   e. Don't overuse a single skill.
   f. Don't use a limited range of skills.

14.1

Activity Unit

Working with two others who have also read this book, role play interviews alternating the client, interviewer, and observer roles. Record each interview on audio- or videotape. A 35- to 45-minute interview is required for this exercise.

Interviewer role   Practice using appropriate nonverbal behavior throughout the interview. Also practice the verbal skills for rapport building and information seeking, and change skills whenever appropriate in response to the client. Single or combination responses are appropriate in this exercise.

Client role Be cooperative and provide sufficient information for the interviewer to practice. Clients may want to role play the problem of a friend or family member or to enact one of the following roles:

1. An active individual who has just had a heart attack and has been told to slow down.
2. A new employee who cannot understand his new job and who is having difficulty making friends on the job.
3. An individual who is in considerable debt but feels his or her present lifestyle must be maintained in order to keep his or her job.
4. A young adult who is well aware of the importance of education but is bored and uninterested in school.

Observer role Operate the equipment and complete the Practice Interview Checklist (see Table 14.1).

Postinterview discussion Using the taped interview and the Practice Interview Checklist data, the participants should discuss the performance of the interviewer. All participants can learn from this discussion, which should be nonjudgmental, focus on positive as well as less effective responses, and stimulate improved performance for all.

14.2

Activity Unit

Using the tape of the interview produced in Activity Unit 14.1 or the tape of another interview, enter the five least effective interview responses below. Then formulate a better response to replace each original one in the space provided.

Original response 1 _____

_____

_____

Better response _____

_____

_____

Original response 2 _____

_____

_____

Better response _____

_____

_____

Table 14.1

Practice Interview Checklist

Activity	\multicolumn{18}{c}{Interviewer response[a]}
	1	2	3	4	5	6	7	8	9	10	11	12	13	14	15	16	17	18
Effective nonverbal behavior																		
On topic																		
Open question																		
Closed question																		
Minimal encouragement																		
Paraphrasing																		
Summarizing																		
Reflecting feeling																		
Structuring																		
Communicating feeling																		
Communicating immediacy																		
Confronting																		
Self-disclosing																		
Interpreting																		
Information giving																		
Enlisting cooperation																		
Effective response[b]																		

[a]Check the activities that characterize each response.
[b]Indicate the degree of effectiveness on a scale from 1 to 5 in which 1 is ineffective and 5 is very effective.

Original response 3 _____

_____

_____

Better response _____

_____

Original response 4 _____

_____

_____

Better response _____

_____

_____

Original response 5 _____

_____

_____

Better response _____

_____

_____

Cultural Considerations

1. Depending on the client's culture, it is necessary for the interviewer to integrate the skills in this book in ways that respect cultural differences.
2. To interview clients from other cultures, a knowledge of the client's culture, as well as one's own, is necessary.

InfoTrack® College Edition Keyword Search Terms

Counseling Skills

Final Thoughts

This book has been based on the single-skills approach to interviewing. In developing your interviewing skills, you began with the single skill of attending behavior and have gradually gained a repertoire of basic listening and action skills. How you use these skills will influence the way a client thinks about, interprets, and resolves issues and problems. Although you now have a repertoire of skills with which to begin interviewing, remember that mastering these skills is only the beginning of the interviewing process. The client, the context, the nature of the issue or problem, and gender and cultural issues will be important determinants of when and how you apply your skills. Remember, too, that the ability to seek consultation and supervision is the hallmark of a good interviewer, no matter how experienced and skilled. This book has provided you with some beginning ground rules for interviewing and a foundation for your future growth. The rest is up to you.

Additional Resources

Interviewing involves a delicate interplay between a knowledge of "why" you are doing what you are doing and "how" to do it. The primary focus of this book is on how to interview. Instructors may wish to use this book in conjunction with any of the books referenced in Chapter 1 or with any of the following books, which focus on the theoretical aspects of interviewing. Readers may wish to supplement the information they have obtained in this book by consulting some of these books.

Brammer, L. M., Abrego, P., & Shostrum, E. (1993). *Therapeutic counseling and psychotherapy* (6th ed.). Upper Saddle River, NJ: Prentice Hall.

Corey, G. (2001). *Theory and practice of counseling and psychotherapy* (6th ed.). Belmont, CA: Wadsworth.

Cormier, S., & Nurius, P. (2003). *Interviewing and change strategies for helpers: Fundamental skills and cognitive behavioral interventions* (5th ed.). Belmont, CA: Wadsworth.

Egan, G. (2002). *The skilled helper: A problem-management and opportunity-development approach to helping* (7th ed.). Belmont, CA: Wadsworth.

Hill, C. A., & O'Brien, K. M. (1999). *Helping skills: Facilitating exploration, insight, and action.* Washington, DC: American Psychological Association.

Hutchins, D. E., & Vaught, C. C. (1997). *Helping relationships and strategies* (3rd ed.). Belmont, CA: Wadsworth.

Ivey, A. E., & Ivey, M. B. (2003). *Intentional interviewing and counseling: Facilitating client development in a multicultural society* (5th ed.). Pacific Grove, CA: Brooks/Cole.

Neukrug, E. S. (2002). *Skills and techniques for human service professionals: Counseling environment, helping skills, treatment issues.* Belmont, CA: Wadsworth.

Okun, B. F. (2002). *Effective helping: Interviewing and counseling techniques* (6th ed.). Belmont, CA: Wadsworth.

Sue, D. W., Ivey, A. R., & Pedersen, P. B. (1996). *A theory of multicultural counseling and therapy.* Pacific Grove, CA: Brooks/Cole.

Index

Credits